**Please check all items for damages
before leaving the Library.
Thereafter you will be held
responsible for all injuries
to items beyond reasonable wear.**

Helen M. Plum Memorial Library

Lombard, Illinois

A daily fine will be charged for
overdue materials.

APR   2007

The Science of Disorder

The Science of Disorder

Understanding the Complexity,
Uncertainty, and Pollution in Our World

Jack Hokikian, Ph.D.

Los Feliz Publishing - Los Angeles

Copyright © 2002 by Jack Hokikian

For information about permission to reproduce selections from this book, write to
Permissions, Los Feliz Publishing,
P.O. Box 291899, Los Angeles, CA 90029-1899

The text in this book is composed in Palatino
Book design by A. Huertas
Cover design by Lightbourne, copyright © 2002
Printed by Thomson-Shore on permanent acid-free paper

Publisher's Cataloging-in-Publication
(Provided by Quality Books, Inc.)

Hokikian, Jack.
The science of disorder : understanding the
complexity, uncertainty, and pollution in our world /
Jack Hokikian. -- 1st ed.
303 p. cm.
Includes bibliographical references and index.
LCCN: 2001117941
ISBN: 0-9708953-2-1

1. Thermodynamics--Philosophy. 2. Entropy--
Philosophy. 9. Environmental sciences--Philosophy.
4. Human ecology--Philosophy. 5. Order (Philosophy)
I. Title.

QC311.2.H65 2002   536'.701
                          QBI01-200767

Los Feliz Publishing
P.O. Box 291899, Los Angeles, CA 90029-1899

1 2 3 4 5 6 7 8 9 0

*To my mother and the memory of my father,
with love and appreciation*

Contents

Acknowledgments

I am thankful to Harriet H. Forster, my dissertation chairperson, and to C. C. Kim who read the first draft of this book and offered advice and encouragement, which I took very much to heart. I am indebted to Andres Huertas, Thomas R. Hill, and Ron Kenner for going over successive versions of the manuscript and for spending countless hours with me in critiquing its various aspects. Finally, I am grateful to freelance editor Linda Starke for her tactful guidance during the editing process. The remaining disorders are all mine.

Introduction

Our world is changing rapidly and becoming more complex, disordered, and polluted. Economies are in disarray, crimes and conflicts are all too frequent, weapons are multiplying and spreading, and our lives are becoming ever more hectic and uncertain. Despite sustained efforts to control pollution, toxic gases are still spewing into our air, poisonous chemicals and nuclear waste products are piling up, fertilizers and pesticides are contaminating our water and food supplies, and our ecosystem is in a state of distress. And in spite of prolonged attempts to eradicate undesirable viruses, bacteria, and insects, the bugs are thriving and attacking us with greater impunity and intensity.

The explanation to our puzzling situation can be found in the discipline of thermodynamics. The First and Second Laws of Thermodynamics govern all processes and activities—from physical, chemical, and biological to economic, social, and intellectual. For historical reasons, they are called the Laws of Thermodynamics because they were discovered while studying heat and heat engines; yet they are extremely general Laws of Nature, vitally important to humankind.

This book—written from a physicist's standpoint—describes how the Laws of Thermodynamics apply to all areas of human endeavor, including chemistry, cosmology, medicine, education, agriculture, economics, technology, and ecology.

Once we grasp the essence of the Laws of Thermodynamics, we understand why our problems, from economic to environmental to social, are becoming increasingly intricate worldwide. Knowledge of these laws will help us comprehend how Nature works, and thus let us live in harmony with Nature and each other.

Solutions to our problems look very different depending upon whether we neglect or work within the Laws of Thermodynamics.

Thus it is important that we be aware of these laws and their effects on our lives, our society, and our environment so we can formulate policies that are not on a collision course with Nature's Laws.

Before human knowledge disintegrated into a myriad of specialties, science and philosophy were unified in their goal: to discover the truth about how Nature works and to arrive at fundamental Natural Laws through which we can present a unified view of Nature and our place in it. What we are left with today is philosophy without science, while science has been taken over by technology—applied science. As one scientist proclaimed: "Listen to the technology and find out what it is telling you."[1] But technology does not provide us with any principles of Nature, principles needed to derive a practical philosophy of life. For this we have to turn to the Laws of Thermodynamics.

Technology promised us a new world in which we would achieve control over our environment and the forces of Nature, gain access to unlimited and cheap energy and natural resources, and enjoy a youthful life free of diseases and health disorders. These promises have not materialized, and for good reason—because we live in a thermodynamic universe, one that cannot be controlled by our technologies but instead is controlled by the Laws of Thermodynamics.

Although these laws were discovered a century and a half ago, most people—including the educated—do not know enough about them. For example, in *Earth in the Balance*, Al Gore refers briefly to the First Law of Thermodynamics.[2] But he makes no mention of the Second Law, the one that affects us most.

The Laws of Thermodynamics—especially the Second Law—are frequently misunderstood and misrepresented in the literature, which is a serious disorder in human knowledge. Throughout this book, many popular misunderstandings of the Second Law are pointed out and corrected, including the widely held misconception that evolution violates it.

The main ideas of the First and Second Laws of Thermodynamics are explained in general terms in the first two chapters, along with some historical background. They provide the scientific foundation for the book's thesis.

In today's fast-changing environment, where technologies, theories, ideologies, fashions, and information quickly become obsolete and are discarded, the Laws of Thermodynamics provide a good sense of stability and continuity. Not only have they withstood the test of time, they have become stronger and stronger as more and more supporting data have accumulated.

The Laws of Thermodynamics are based on two thermodynamic quantities: energy and entropy.

The First Law is about the conservation of energy. It says the amount of energy in the universe is constant. This implies that energy cannot be created or destroyed but can be transformed from one form to another. The expression "You can't get something for nothing" stems from this law.

The Second Law is about entropy. It stipulates that entropy increases in all processes irreversibly. Physicists identify entropy as a measure of the disorder of a thermodynamic system. In economic terms, the Second Law can be regarded as Nature's unyielding tax collector. It exacts a tax from all our activities by increasing the disorder of our thermodynamic system. Through increases in entropy, the Second Law controls and dictates the way all processes proceed in the universe. For this reason, it maintains a supreme position within the Laws of Nature. It demands our undivided consideration.

Whether we are physicists, biologists, economists, psychologists, or politicians; whether we are conservatives, liberals, or middle-of-the-roaders; whether we are technologists, environmentalists, deep ecologists, or ecofeminists; whether we are Africans, Americans, Asians, Australians, or Europeans; whether we are capitalists, communists, or socialists—we all feel and are affected by the cumulative effects of the physical, social, environmental, economic, and intellectual entropies within us and around us. Consequently, it is to our advantage to learn and understand what entropy is all about.

Three decades ago, ecologist René Dubos remarked that "the technological and other practical applications of science have been oversold." This eminent biologist believed that science would be more useful to humanity if it devoted more energy toward "the development of knowledge and attitudes that would help man to examine objectively, rationally, and creatively the problems that are emerging as a result of social evolution." He pointed out that "this aspect of science is given very low priority—if not neglected altogether—in universities and research institutes," adding that "we hardly give any thought to the long-range consequences of our scientific and technological interventions into man's life and nature."[3] Dubos' observations are as relevant today as they were then.

Regrettably, the Laws of Thermodynamics have not received the emphasis they deserve. They can help us understand and tackle today's problems, particularly problems of our own making. Moreover, the discipline of thermodynamics includes important universal concepts, such as the irreversibility of natural processes. Our perspective on the world is very different if we view it as a reversible system subject to our control rather than an irreversible system governed by the Laws of Thermodynamics.

Once we become familiar with these laws, many previously unexplained phenomena and paradoxes become apparent. We see why time flows only one way, why we age irreversibly, why our lives are becoming increasingly complicated and uncertain, why we are experiencing "future shock," why we have less time for ourselves even though we are surrounded with more "timesaving" devices than ever before. We realize why so many promises and expectations have not come true. We are also able to foresee some dangers ahead.

The current educational environment and economic conditions have created a lot of specialists. Yet the concepts we have learned in our specialties, while useful and functional, have narrow ranges of applicability. They do not equip us with the ability to view the world in a general, comprehensive way. On the other hand, the discipline of thermodynamics—through its all-encompassing laws—allows us to see the whole picture. Indeed, it forces us to examine the total picture. Many of our gross errors in judgment have come about because we have considered only part of the thermodynamic system.

To deal with today's diverse but interconnected set of problems, we need a common set of general principles of Nature, principles that apply to all processes and activities. Then we will have a basis for discussing and tackling our pressing problems in economics, government, education, health care, transportation, technology, and ecology.

Chapter 1

Nature's First Law

Of all the concepts or constructs of physics, energy, by its unifying capacity, has proved by all odds to be the most significant and successful. Its domain of application has indeed by now far transcended physics and covers all branches of science. . . . it is the physical construct which has proved to contain the greatest meaning for all aspects of human life.[1]

—R. Bruce Lindsay

1

What is Energy?

Energy is "capacity to do work." Whether we are preparing breakfast, jogging, or building a house, we need energy to perform activities—all different manifestations of a concept called work. Without energy, no work can be done. Of course, our ancestors were unaware of the equations of physics that link energy with work; nevertheless, using materials that Nature provided, they spent energy to build tools that in turn performed useful work for them.

In physics, "work" has a special meaning. We feel we have done twice as much physical work when we push a couch 10 feet rather than 5. Thus "work" is defined as force applied in the direction of the body's movement times the distance the body moves.

Physicists and nonphysicists have different perceptions of what work means. Suppose we see a big piece of rock and try to move it with our bare hands. We might spend considerable effort, perspiring a great deal, but the rock may be too heavy and not move at all. Undoubtedly this is considered hard work. According to a physicist, however, we have not done any work, since the rock did not move: there was no displacement. We complain that such an equation of work is unfair.

"Wait a minute," says the physicist, "you definitely spent appreciable energy, but you did not convert the dissipated energy into mechanical work." We are reminded that energy is capacity to do work. Energy is required to perform mechanical work. But it is possible to dissipate energy and not perform work in the technical sense—for example, holding a heavy suitcase in a stationary position.

Nature undergoes a multitude of changes, but scientists look especially for Nature's constants. In 1669, Christian Huygens, while looking for such constants, observed that the quantity mv^2—mass of a moving object multiplied by the square of its velocity—remains unchanged in perfectly elastic collisions, such as collisions between steel balls. Baron G. W. von Leibnitz later named it the living essence of a moving body, or its *vis viva*.[2]

However, if we throw a stone up in the air, we notice that it gradually loses velocity, comes to a stop and starts to come down—all the while increasing speed until it hits the ground. In this case, the living essence of a body does not remain constant.

Gabrielle du Châtelet, a lady of the French court, made her own proposal. (In the eighteenth century, the mechanistic worldview had picked up momentum, and it became fashionable for ladies of the court to dabble in Newtonian physics.) She suggested that the living essence could be transformed to *vis mortua*, or dead essence. She had the vision of a conservation law involving the two quantities—*vis viva* and *vis mortua*—that could be converted into each other but the sum of which remained constant.[3]

Gradually the term "energy" began to replace the words *vis viva* and *vis mortua*. *Vis viva* became actual energy and *vis mortua* became potential energy. Later physicists halved the quantity mv^2 and called the result kinetic energy ($mv^2/2$). Finally, with the help of Joseph Lagrange's theoretical work, it was possible to enunciate that total mechanical energy, which is composed of kinetic and potential energy, is constant in an isolated system of bodies.[4] Applied to the solar system, the law was an instant success. Earth, in its elliptical orbit, gains velocity as it moves closer to the Sun, and slows down as its distance increases. While Earth's kinetic and potential energies change in value during the orbit, their sum remains constant.

But serious complications arose right here on Earth. We notice that all moving objects come to a stop if left alone. A swinging pendulum will exchange potential energy with kinetic energy, back and forth, but eventually—due to frictional forces—will cease to move. Unfortunately, we cannot neglect friction when dealing with physical systems. The pendulum is slowed by frictional forces, which cause energy dissipation. As it moves, the pendulum must do work against the frictional forces. This extra work is the same amount as the energy dissipated.

The gravitational potential energy of an object is defined by its mass times the downward acceleration due to Earth's gravity, multiplied by the height of the object above the ground. One of the oldest methods of harnessing energy is to take advantage of gravitational potential energy owing to differences in water levels. This kind of energy is called hydropower, in use at least since the first century B.C. to grind grain.

The Emergence of the First Law of Thermodynamics

The science of thermodynamics introduces a new concept, that of temperature.[5]
—Arnold Sommerfeld

In the natural sciences, heat (temperature) is an important concept because heat is somehow involved in all natural processes. But what is heat? Two conflicting views of the nature of heat persisted for centuries: the materialistic and the motional. The materialistic view held that heat consisted of a "real, material substance, a very subtle fluid" capable of penetrating all space and able to flow in and out of all objects.[6] Presumably, the presence of this "imponderable matter of heat" made objects hot, while its absence made them cold.[7] In 1789, this "elastic fluid that produces heat" was named caloric by Antoine Lavoisier and his colleagues.[8] The adherents of the caloric theory included such scientific giants as John Dalton and Joseph Priestley.

Conversely, Francis Bacon was convinced that heat is due to the motion of the particles of matter, but his views were not clearly expressed. Robert Boyle was a more articulate defender of the motional theory of heat, and offered as well a definite experimental reason for his thesis. Boyle pointed out that a nail becomes hot when hammered into a block of wood, declaring that heat generated by mechanical means is "new" heat.[9]

In 1798, Benjamin Thompson (Count Rumford) made a similar discovery. While working on cannons for the government of Bavaria, he noticed that a great deal of frictional heat is generated in boring a cannon. Impressed by the phenomenon, Thompson then devised an ingenious experiment. Deliberately using blunt borers, he observed that they cut very little metal but produced just as much heat as sharp borers. He became suspicious of the validity of the caloric theory.

One morning, just after breakfast, Thompson immersed both cannon blank and borer in a big tank of water and began boring. The instrument was so dull that very little was cut by lunchtime, and yet enough heat was generated to boil the water. After a long Bavarian lunch the water cooled, but redrilling boiled the water again. He repeated the experiment with the same results. He became convinced that friction and work are convertible into heat and that heat cannot possibly be anything, "except it be MOTION."[10]

Thompson's experiments provided strong evidence of the validity of the motional theory. By the end of the eighteenth century, however, the caloric theory was thoroughly entrenched and firmly held. Evidently, the time was not ripe for the fusion of the concepts of mechanical energy (work) and heat into a useful generalization. We had to wait until the work of Mayer, Joule, and Helmholtz some four decades later.

Whenever the First Law of Thermodynamics is mentioned, we tend to think of physics and heat engines. Interestingly, the principle was first enunciated not by a physicist but by a physician from Heilbronn, Germany—Julius Robert von Mayer (1814–78). He had an inquiring mind, which pushed his thoughts beyond the confines of medicine.

In 1840, Mayer was engaged as a ship's doctor on a voyage from Rotterdam to Surabaja in the East Indies. By the time the ship reached its goal, some sailors had come down with fever. While bleeding them, he noticed the color of the blood taken from their veins was much brighter than what he was used to seeing in Germany. This seemingly insignificant difference intrigued him. While other members of the crew got off the ship to explore the Indies, Mayer stayed on board and pondered what he had observed. His curiosity led him to discover one of the most basic and useful laws of Nature.

Mayer reasoned that in the hot tropics the body did not need to burn up as much energy to maintain its temperature as it did in cold Germany. The blood's bright color, then, was due to less oxidation.

Here he was very close to the central idea of the First Law of Thermodynamics: the amount of energy in the universe is constant; energy cannot be created or destroyed, but it can be converted either to heat or to work, or both. Work and heat must therefore be equivalent concepts, and mutually convertible in some ratio.

Thus Mayer began his pursuit of the equivalence between mechanical work and heat. Because he also was interested in philosophical considerations of cause and effect in natural phenomena, he asked: What is the origin of heat? This question brought him closer to physics. His unfamiliarity with the subject was an asset at first, but became a serious problem when he sought to make his theories known and accepted.

Because Mayer was not acquainted with the prevailing caloric theory, he innocently jumped to the right conclusion that friction, percussion, or any sort of mechanical work can produce heat. After returning to Germany and gaining some familiarity with physics, he went to physicist Philipp von Jolly for consultation. Mayer tried to explain that heat is due to motion, and that heat and work are equivalent concepts. However, Jolly had great difficulty understanding him, and finally exclaimed: "But if what you say is true, then water should be warmed by merely shaking it." Mayer left quietly, only to return several weeks later to Jolly's room, exclaiming, "And so it is."[11]

Mayer tried to publish his discoveries and ideas. He submitted a paper to the prestigious *Annalen der Physik*, which refused publication. Although the ideas were correct, the language was blemished by some serious errors in terminology. For example, he used the term "force" (German *Kraft*) where he meant energy. Mayer corrected some of his errors and sent a short account of his theories to the *Annalen der Chemie und Pharmacie*. The paper was published in 1842, but received little attention. The introduction was not very scientific, containing a metaphysical dissertation that repelled scientists. Old Latin formulas like *ex nihilo nihil fit* (from nothing nothing is produced) were used to prove that work and heat are equivalent and mutually convertible, and that the force they represent is indestructible. There is no doubt, however, what Mayer meant: energy is always conserved.

If mechanical work and heat are equivalent, then what is the numerical equivalence? No one had made direct measurements or calculations to determine the quantitative relation between work and heat. In the last paragraph of his paper, Mayer gives the value of what is now called the "mechanical equivalent of heat." He does not give a clear and detailed account of how he obtained it; nevertheless, modern historians find his paper "epoch-making."[12]

Although discouraged by the lack of attention, Mayer did not give up. In 1845, he expanded his theories in a pamphlet, which he printed at his own expense. There he gave a detailed account of how he had calculated the mechanical equivalent of heat. His approach was truly innovative and ingenious. Because he could not afford a fancy laboratory, he had gone to the literature and found a long-forgotten experiment by Joseph Louis Gay-Lussac. It was from Gay-Lussac's data of the free expansion of gases that Mayer had calculated the mechanical equivalent of heat. Unfortunately, one of the physical constants (specific heat of air at constant volume) he used was not then known accurately. Thus despite his clever and correct method of interpreting the data, his computation was off.

In his pamphlet, Mayer also gave an account of his extensive investigations of the food consumed and the work done by humans and animals. He extended his theories to the organic world. He stated that mechanical work, heat, chemical action, light, electricity, and magnetism are all different forms of force (energy), mutually convertible and equivalent, and that the totality of these forces is unalterable in the whole universe.[13] In one shot, Mayer accomplished the following: He equated heat to mechanical work, thus rejecting the accepted caloric theory; demonstrated that energy principles apply to both the living and the nonliving; and elevated the energy conservation principle to a universal law.

Although Mayer's ideas were correct and considerably advanced, scientists ignored them and the local press ridiculed them. In addition, Mayer watched others publish essentially the same results and receive credit for their work. One night in 1850, suffering intensely from the neglect of his work, Mayer threw himself out of a window to the paved street two stories below. The fall was not fatal, but it added substantial physical pain to his mental anguish.

The second independent contributor to the First Law of Thermodynamics was an amateur scientist who achieved genuine success and renown in his chosen field of physics. Born into a rich family of brewers at Sanford near Manchester, England, James Prescott Joule (1818–89) had the independent means to devote all his time to research on scientific subjects that interested him. Largely self-taught, and informally tutored by Dalton, he became interested in the relatively new science of electromagnetism. At the age of 19, he devoted much time to the development of electric motors, with the idea that they would prove more efficient than steam engines. He was soon disabused of this hope.

Joule's work was not wasted, however, because he learned that heat always plays a role in the production of mechanical effects by electricity. His early experiments on the production of heat by electric currents made him skeptical of the caloric theory. So he devised a simple but straightforward apparatus—the paddle-wheel box. He became famous for this invention because the experiment was instrumental in overthrowing the caloric theory and achieving a tremendous task: the establishment of the First Law of Thermodynamics. Appropriately, a modern energy unit, the joule, is named after him.

In his paddle-wheel experiment, Joule immersed a thermometer in an insulated tank full of water, tied a heavy weight to a string that went over two pulleys, and attached the other side of the string to a paddle wheel submerged in the water. He then released the weight at the top. The weight fell very slowly, gaining almost no velocity, hence no kinetic energy. The potential energy was converted to mechanical energy (work), which rotated the paddle wheel, stirring the water and raising its temperature. In the process, mechanical work was transformed to heat (thermal energy).

This was a landmark experiment. Joule had expressed the mechanical value of heat unambiguously in terms of work, experimentally verifying Mayer's theoretical determination. From a great number of these experiments, Joule concluded, in today's parlance, that it takes 4.154 joules of mechanical work to get one calorie (c) of heat. Despite all the technological advances since then, nobody has improved his value by more than 1 percent, a truly remarkable achievement. Today's accepted value is 4.186 joules for one calorie.

A calorie is defined as the quantity of heat needed to raise the temperature of one gram of water by 1 degree Celsius. This is called the small calorie (c); it is not the one we count when we go on a diet. Food calorie is the big Calorie (C), equal to 1,000 small calories. One Calorie, then, corresponds to 4,186 joules of mechanical work, and one joule of energy is equivalent approximately to lifting one tenth of a kilogram of weight a distance of one meter. This is why it is so difficult to lose weight by exercise alone.

Although Joule was a superb experimentalist, he attracted little attention from the scientific community. In 1847, after further refining his experiments, Joule went to the British Association meeting at Oxford. There, the chairman suggested that he give only a short verbal account of his paper, since there were so many papers scheduled. His work was just about to go unnoticed one more time, when a young man stood up and by his sharp questions started a lively discussion. That man was William Thomson (Lord Kelvin), who stayed after the meeting and had a long conversation with Joule, beginning a relationship that "quickly ripened into a life-long friendship."[14]

The great scientists attending that meeting were not convinced of Joule's bold conclusions, since he had to rely on only small fractions of a degree to prove his case. Later, Lord Kelvin recalled that he was "tremendously struck" by Joule's paper.[15] Finally, in 1849, Joule received the seal of approval when his paper "On the Mechanical Equivalent of Heat" was communicated to the Royal Society by the all-around scientist Michael Faraday. In 1870, Joule was awarded the Copley Medal of the Royal Society, its highest honor.

Like Mayer, Joule had to struggle for recognition. He was more fortunate, however, as he had more money and an excellent facility for experimental work. Unlike Mayer, Joule succeeded in overthrowing the accepted orthodoxy, the caloric theory, and in establishing the First Law of Thermodynamics, thanks to his "great experimental skill, courageous persistence and great good luck in meeting Kelvin at the right time."[16]

Impossibility of Perpetual Motion Machines

From Monte Carlo to Las Vegas to Atlantic City, thousands of people, rich and poor, gather around roulette wheels to try their luck. Some win, some lose. Those who lose can blame probability theory or temporary bad fortune. But those who win should be thankful to the French philosopher and mathematician Blaise Pascal (1623–62), who in his quest for a perpetual motion machine invented the roulette wheel and the theory of roulette. (Perpetual motion machines perform work in perpetuity without any input energy.)

Of course in the seventeenth century, the conservation of energy law—which dispels the possibility of perpetual motion—had not yet been discovered. But in 1847, Hermann von Helmholtz derived the principle of energy conservation by using the fact that all efforts to produce perpetual motion machines had failed.[17]

Helmholtz, the third independent discoverer of the principle of the conservation of energy, was a philosopher and a man of science. Since his parents were poor and felt that a purely scientific life could not be sustained, he became a physician and surgeon, and taught physiology at a number of German universities.

In 1847, at the age of 26, Helmholtz presented a paper before the Berlin Physical Society on the conservation principle. Although Helmholtz showed detailed knowledge of physics, he attracted little attention. And what little attention he did receive was mainly hostile. His paper to *Annalen der Physik* got the same treatment as Mayer's; it was rejected. Like Mayer, he published the paper privately in a pamphlet form.

Helmholtz could not prove his fundamental axiom that energy is conserved. So he used the reverse process. He pointed out, correctly, that all attempts to produce a perpetual motion machine had failed and that physicists were in general convinced of its impossibility. He also emphasized that Sadi Carnot and Emile Clapeyron had used essentially the same argument and deduced important laws of heat, which were later verified experimentally. As a final support for his case, Helmholtz remarked that as early as 1775 the Paris Academy of Sciences decided to accept no more claims for perpetual motion machines, putting them in the same category as circle-squaring.[18]

Helmholtz showed that heat and work are different forms of energy, and that their sum is always conserved. This is the essence of the First Law of Thermodynamics, the principle of the conservation of energy applied to phenomena involving work and heat. The First Law does not measure energy but the changes—increases or decreases—in the internal energy of a thermodynamic system. The law states that the change in the internal energy (ΔE) of a system is equal to the sum of the heat (Q) absorbed and the work (W) expended in producing the change ($\Delta E = Q + W$).

Mechanical work is measured in terms of force times distance, while heat is measured in calories. Two physical quantities cannot be added together unless they have the same units or their equivalence is quantitatively known. Thus Helmholtz was enchanted when he first heard that Joule had measured the mechanical equivalent of heat. He quickly used it in support of his conservation of energy principle and showed in clear, scientific language that Mayer's calculations were in agreement with Joule's experimental results.

Physics is often involved in the pursuit of unknowns. To solve a problem containing several physical unknowns, physicists need as many independent facts (physical equations) as unknowns. Otherwise, the problem remains impossible to solve. In physics, the energy conservation principle often provides the equation that unravels the entire puzzle.

Helmholtz used the energy conservation equation exactly for this purpose. He solved physics problems that could not otherwise be solved, demonstrated with many examples the theoretical importance of the law and its practical usefulness, and suggested that the foremost task of physics should be the full verification of the law.

At the beginning, Helmholtz's arguments fell onto deaf ears. He was not an influential physicist then, but his suggested path for physics was the right one. Subsequently, physicists did precisely what Helmholtz expressed. Today they view him as one of the top theoretical physicists who ever lived.

In 1871, Helmholtz was appointed professor of physics at the University of Berlin. (He became Max Planck's teacher.) He died in 1894 after making many valuable contributions to physiology, anatomy, and several branches of physics.

The ideas and methods of Mayer, Joule, and Helmholtz gradually won acceptance in the scientific community. Joule and Helmholtz did not receive instant recognition, but were not ridiculed like Mayer who, after his suicide attempt, went mad and was committed to an asylum where he was badly treated. Two years later, Mayer was released, but he never quite recovered his mental stability. He dropped out of the scientific scene for a decade, during which time he is reputed to have cultivated his vineyard at Heilbronn.

Some artists and scientists are glorified after their death. In Mayer's case, he was scorned by his peers, laughed at, and harshly treated during his productive years, but received acknowledgments and praise during the latter part of his life. The first to recognize Mayer's work was Helmholtz. In 1854, at a popular lecture attended by many physicists, Helmholtz described Mayer's pioneering work and acknowledged that Mayer was the original propounder of the principle of energy conservation.

Physicists John Tyndall in England and Rudolf Clausius in Germany repeated Mayer's calculations, using the latest specific heat measurements of Victor Regnault. They found Mayer's work in full agreement with Joule's experimental result.

Mayer's misfortunes began to change. Tyndall accepted his contributions as genuine. In 1862, he delivered a lecture at the Royal Institution devoted entirely to Mayer's work. Clausius wrote a letter to Mayer, praising his work and telling him about Tyndall's lecture. The news that Mayer's contributions had been approved in England spread quickly in Germany. The man who had been ignored, ridiculed, and abused finally started to receive the recognition and honors he so much deserved. Mayer received the Poncelet Prize, and the King of Württemberg ennobled him. In 1871, the Royal Society awarded him the Copley Medal. The tormented "prophet of energy" died a revered old man in 1878.[19]

According to the First Law, a thermodynamic body can raise its internal energy only if the overall energy (heat + work) it receives from its surroundings is positive. The machines we build perform work on their surroundings, and therefore work is negative in the equation. If a thermodynamic system absorbs no energy or does no work on its surroundings, its internal energy remains constant. This is the principle of conservation of energy. The thermodynamic system can be a physical, chemical, biological, or any other kind of system. There is no restriction as to the kind of body that the First Law applies to; everything is a thermodynamic system, from a Thermos jug, to a human body, to a house, to a chemical plant, to the entire universe. Because of this all-unifying power, the First Law of Thermodynamics has been called the greatest generalization of the natural sciences.

Since the days of Mayer, Joule, and Helmholtz, new fields of physics have sprung up. Physics has become more abstract and more mathematical—in some cases abstruse—but the verification of the First Law has remained essentially the same. It is based, as German physicist Max Planck put it, on "the fact which has been tested by centuries of human experience, and repeatedly verified, viz. that *it is in no way possible . . . to obtain perpetual motion, i.e.* it is impossible to construct an engine which will work in a cycle and produce continuous work, or kinetic energy, from nothing."[20]

Perpetual machines have always been humanity's dream. To get something from nothing seems to be our eternal wish. Even now, somebody, somewhere, is working on such an invention. Perpetual machines are a symbol of permanence and eternity. They give us the illusion of no beginning and no end.

The First Law gives us a framework for what we can do and what we cannot do. We can transform energy from one form to another, but we cannot create energy or destroy it. The purpose of discovering natural laws is to help us eliminate the impossible, thereby saving us a great deal of useless effort.

Radioactivity and Perpetual Energy

At the close of the nineteenth century, just when physicists were feeling comfortable with the First Law of Thermodynamics, the French physicist Antoine Henri Becquerel discovered radioactivity. This potentially infinite source of energy created an intellectual and scientific "energy crisis." It seemed physicists had finally found the source of perpetual energy. The much-tested principle of conservation of energy, which had permeated all branches of physics and natural sciences, was in deep trouble. Physicists had some work to do.

Within a few years, Pierre and Marie Curie began experiments with radioactivity and found that pitchblende, a uranium ore, was more intensely radioactive than pure uranium. After some chemical separation of uranium, intense radioactivity remained. That prompted the Curies to undertake further chemical separation of uranium. In the process, they discovered two distinct radioactive elements, which were later named polonium and radium.

The Curies' discoveries were exciting, but they created a serious problem for physicists. The "inexhaustible" energy emitted by radioactivity seemed to violate the existing framework of the principle of conservation of energy. However, the observed radioactivity did not turn out to be a source of perpetual energy. Physicists noticed that the intensity of polonium's radioactivity decreased—in a curious way—over a period of months. Approximately every 140 days, the intensity was halved. So 280 days from the start of the experiment, the intensity was one fourth of the original, and this degradation continued. All radioactive materials decay in this fashion, although the time scale varies from one case to another. Physicists refer to this "halving-time" as the half-life. The half-life of radium is about 1,600 years, whereas that of uranium is 4.5 billion years, the approximate age of Earth.

Although radioactivity was discovered experimentally by Becquerel and the Curies, the theoretical foundation of nuclear energy came from Albert Einstein. In 1905, he stated that mass is a form of energy. His famous equation $E = mc^2$ (energy is equal to mass times the square of the speed of light) started a new era for physics and humankind.

Einstein's equation originated from his theory of relativity. Although theoretically correct, the equation would be more meaningful as $E = \Delta mc^2$, where Δm means change (loss) in mass. As Edward Teller points out, if Einstein's equation were achievable in practice energy would not be a problem. Since c^2 is a very large number ($c = 186,000$ miles per second), "a ton of any material would supply the energy needs of the United States for a year."[21] As it turns out, we can only convert certain kinds of matter into energy, and then only in small fractions.

Nuclear physicists began to use Einstein's equation to test the law of conservation of energy. Precise measurements showed that the conservation law indeed holds in radioactivity and nuclear processes; the measured masses and energies were adding up precisely. Today, thanks to an overwhelming amount of experimental evidence, the First Law of Thermodynamics is no longer questioned in any branch of the natural sciences, including the microscopic world of nuclear and particle physics.

Chapter 2

Nature's Second Law

A good many times I have been present at gatherings of people who, by the standards of the traditional culture, are thought highly educated and who have with considerable gusto been expressing their incredulity at the illiteracy of scientists. Once or twice I have been provoked and have asked the company how many of them could describe the Second Law of Thermodynamics. The response was cold: it was also negative. Yet I was asking something which is about the scientific equivalent of: Have you read a work of Shakespeare's?[1]

—Sir C. P. Snow, *The Two Cultures*

The Beginning of a New Science

While James Watt's invention of the steam engine in 1765 is the landmark symbolizing the beginning of the industrial revolution, the work of the French military engineer Sadi Carnot set the stage for understanding how steam engines really work.

Nicolas Léonard Sadi Carnot was the eldest son of Lazare Carnot, Napoleon's Minister of War. In 1820, at the age of 24, Carnot went on half-pay from the French army to devote his time to studying physics and economics. He made good use of the educational and research facilities of Paris. In 1824, he summarized his findings in a short book, only 118 pages long.[2]

Carnot's seminal work remained unnoticed for about a quarter of a century. In fact, when Lord Kelvin was working in Paris and tried to obtain a copy of the book, he found that the Parisian booksellers had not heard of it or its author.[3] When he finally acquired one, he was impressed by it, considering it a "most valuable contribution to physical science."[4] Its content essentially launched a new theoretical science, which Kelvin called "thermo-dynamics."

Carnot was a master synthesizer. By disregarding the details of how a steam engine operates, he was able to concentrate on its truly significant and general features: A heat engine receives energy—from a hot reservoir—in the form of heat (Q_2) at a relatively high temperature; the thermal energy is used to perform mechanical work; and the engine ejects heat (Q_1)—usually to the environment—at a lower temperature. (The exhaust heat, called waste heat, does not contribute to the production of work.) He came up with an expression for the maximum efficiency of any heat engine operating between two given temperatures: $(Q_2 - Q_1)/Q_2$. Carnot then established this general and far-reaching proposition: *"The motive power of heat is independent of the agents employed to realize it; its quantity is fixed solely by the temperatures of the bodies between which is effected . . . the transfer of the caloric."*[5]

This principle became the catalyst for a more scientific temperature scale than the Fahrenheit and Celsius scales already in use. The Fahrenheit scale was invented in 1714 by German physicist Gabriel Fahrenheit. He set the lowest temperature he could obtain in his laboratory in East Prussia at zero, and set the normal temperature of the human body at 96. In this scale, ordinary water froze at slightly under 32 and boiled at a slightly under 212. So Fahrenheit set the freezing point of water at 32 and the boiling point at 212. The normal body temperature thus became 98.6 degrees Fahrenheit.

In 1742, Swedish astronomer Anders Celsius devised a different temperature scale. In this, 0 degrees is defined as the freezing point of water, and 100 degrees is defined as the boiling point of water, at normal atmospheric pressure. This is known as the Celsius or the centigrade scale, from the Latin meaning "hundred steps."

Lord Kelvin had another idea. He proposed using Carnot's principle to define a new temperature scale, which he called "absolute scale." He had always felt the need to establish a temperature scale "independent of the physical properties of any specific substance."[6] Since the size of the degree is arbitrary, Kelvin chose the centigrade.

Recognizing that heat is disordered motion due to temperature, Kelvin designated the point where all motion comes to rest as 0 degrees. He found that in this new absolute scale, water freezes at 273 degrees (273.16 degrees to be exact) and boils at 373 degrees. The new measure was named the Kelvin scale and is symbolized by K. (Any temperature in Celsius can be converted to the Kelvin scale simply by adding 273.) This new temperature scale is truly thermodynamic in nature and the most fundamental one because it is independent of the properties of any particular substance. Using this scale, Carnot's maximum efficiency principle for a heat engine can be expressed simply in terms of the temperatures involved in the process: $(T_{high} - T_{low})/T_{high}$.

Physics and engineering books are full of equations, but Carnot's maximum efficiency expression truly stands out as one of the most important ones. From this came the Second Law of Thermodynamics, which spread into every field of the natural sciences and then into many other fields of human knowledge.

Carnot's principle gives us a yardstick to measure the economy—efficiency—of heat engines. For example, if a steam engine operates between the boiling and the freezing point of water, its maximum efficiency can be calculated: $(373 - 273)/373 = .268$. Thus, only 26.8 percent of the heat can be converted into useful work, leaving 73.2 percent as simply unusable. This is true for any heat engine operating under ideal conditions between these two temperatures. Carnot's efficiency principle tells us that we need a difference in temperature to extract work from a heat source. Thermal energy, by itself, is not enough.

Until the middle of the nineteenth century, it was thought that with sufficient ingenuity, efficiencies close to 100 percent might be obtained in heat engines. Carnot's efficiency principle changed the attitudes of scientists and engineers. It set a limit to technological capabilities. And in general, technology achieves less than the theoretical limit allows, because of difficult practical considerations.

The thermodynamic limit imposed by Carnot's efficiency equation is highly relevant to our modern life. Most energy supplies that we use to obtain mechanical or electrical energy come in the form of fuels, such as coal, gas, or oil, where the energy is stored as chemical energy. The process of combustion releases the chemical energy and transforms it into thermal energy. In thermal form the energy may be used directly for cooking, heating, or other purposes. To run a vehicle, operate a machine, or generate electricity, however, the thermal energy must be converted to mechanical or electrical energy. One of the responsibilities of the engineer is to carry out this conversion with maximum possible efficiency, subject to practical and economic considerations.

The remarkable achievement of Carnot's work is that he formulated the fundamental statements about energy efficiency before the concept of the energy content of heat was clearly specified. A quarter-century before the First Law was formulated, Carnot had found a fundamental principle that helped scientists discover the Second Law of Thermodynamics. Scientific discoveries and advances do not always follow a logical order.

The Birth of the Second Law of Thermodynamics

Carnot's energy efficiency equation created a scientific problem. Physicists were asking: If all the heat in a heat engine cannot be totally transformed to mechanical work, as Carnot claims, what is then happening to the thermal energy? Is it lost? James Prescott Joule could not accept that. He wrote, "Believing that the power to destroy belongs to the Creator alone I affirm . . . that any theory which, when carried out, demands the annihilation of force, is necessarily erroneous."[7] Ironically, Joule, who spent his whole life as an experimentalist, was challenging Carnot's principle on grounds that lay outside of physics.

The point, however, was valid. According to Joule's experiment, a given amount of work can always be converted to an equivalent amount of heat. Is the reverse true? Kelvin, who had read Carnot's work, did not think so.[8] According to Kelvin's biographer, Silvanus Thompson, "The apparent conflict took possession of [Kelvin's] mind and dominated his thoughts."[9]

Kelvin and Rudolf Clausius of Germany undertook the task of reconciling the work of Carnot and Joule. Their efforts paid off handsomely. In the process, they formalized and developed the First and Second Laws of Thermodynamics that, to this day, remain the Supreme Laws of Nature.

The process of reconciliation began with another question, this time by Kelvin himself: What happens to the mechanical work that is lost when heat flows freely from a hot body to a cold one, not by means of a heat engine but by simple conduction? Kelvin noted that "Nothing can be lost in the operations of nature—no energy can be destroyed."[10] He then asked: What effect is produced in place of the mechanical work that is lost? He called for an explanation.

Scientific endeavors are greatly enhanced by asking the right questions, because someone may eventually come up with the right answers. It took the creative mind of Clausius to come up with an answer to Kelvin's query. He was the first to recognize that Joule's and Carnot's axioms are logically independent; one does not contradict the other. His explanation was straightforward. In a heat engine, heat may be viewed as composed of two parts. A certain portion of heat that is drawn from the hot reservoir may be consumed in producing work, and a further portion transmitted to the cold reservoir. Moreover, he pointed out that the maximum amount of work from a given temperature difference is determined by Carnot's energy efficiency expression. He concluded that both principles "may not only exist together, but that they mutually support each other."[11]

Kelvin felt he now had an answer to Joule's challenge and to his own query. What happens to the mechanical work that could have been harnessed in the case of simple thermal conduction? It is, Kelvin replied, "irrecoverably lost to man, and therefore 'wasted,' although not *annihilated*."[12]

The concept of irrecoverably lost (wasted) energy can be applied to other everyday situations. For example, what happens to the kinetic energy of the water from Niagara Falls if we do not harness it by means of a turbine? The kinetic energy is transformed into heat when the water hits the ground and this energy is irrecoverably lost.

Once Joule's challenge was satisfactorily met, Kelvin and Clausius went to work on further developing and formalizing Joule's and Carnot's axioms. Although they worked independently, they often supplemented and paraphrased each other's statements. Clausius excelled in mathematics and Kelvin in simple, direct statements that could be easily understood. And both, unlike Mayer, Joule, and Helmholtz, had no difficulty whatsoever in publishing their papers. In fact, several of Clausius's papers appeared almost simultaneously in German, English, and French journals—evidence that there are definite advantages to being part of the establishment.

Although Joule's objection to Carnot's axiom had been settled, a fundamental query of natural science still remained unanswered— namely, the question of the direction in which thermal processes occur in Nature. Whether heat flows from a hotter to a colder body or vice versa cannot be answered by the First Law. To resolve the issue, a new principle of thermodynamics was required.

Carnot's work brought to people's attention that in Nature there might be a certain favored direction of energy flow. Clausius added a new fundamental proposition of thermodynamics that *"heat cannot of itself pass from a colder into a warmer body."*[13] That is, heat flows from a warmer to a colder body. This simple statement was the basis of the Second Law of Thermodynamics, which Kelvin and Clausius were the first to formulate: It is impossible to build an engine that operates on a cycle whose net effect is the cooling of a heat-reservoir and the raising of a weight.

This axiom says it is impossible to obtain work by cooling a body below the temperature of its surroundings. Suppose you could cool water and convert the extracted heat into mechanical energy. Then the immense supply of the oceans' thermal energy could be converted into useful mechanical work. For instance, a ship could take in ocean water, convert it into ice, and in so doing propel itself.

Traditionally all machines that have attempted to violate the First Law have been called perpetual motion machines of the first kind. In the same tradition, the German physical chemist Wilhelm Ostwald labeled perpetual machines of the second kind all those that seek to violate the Second Law of Thermodynamics.

As Max Planck pointed out, if such an engine could be built it could be used simultaneously as a motor and a refrigerator without any waste of energy or material; it would be the most profitable device ever made. It would not be perpetual, since it would not produce work from nothing but from the heat drawn from some reservoir. Nevertheless, such a device would give humans the essential advantage of perpetual motion, since there is so much heat in the earth, the atmosphere, and the sea.[14]

Nature's Irreversible Trend

General thermodynamics proceeds from the fact that, as far as we can tell from our experience up to now, all natural processes are irreversible. Hence according to the principles of phenomenology, the general thermodynamics of the second law is formulated in such a way that the unconditional irreversibility of all natural processes is asserted as a so-called axiom.[15]
—Ludwig Boltzmann

Reversible processes are not, in fact, processes at all, they are sequences of states of equilibrium. The processes which we encounter in real life are always irreversible processes.[16]
—Arnold Sommerfeld

The Second Law of Thermodynamics has a much wider meaning than "heat flows from hot to cold." The principle expressly states that it is not possible to reverse completely any process that has occurred in Nature. A process that can in no way be completely reversed is called irreversible. All other processes are called reversible. "Completely" means that the initial state has been exactly restored in its entirety.

Whether reversible processes exist in Nature is not automatically evident. Max Planck asserts that there is only one way of "clearly showing the significance of the second law, and that is to base it on facts by formulating propositions which may be proved or disproved by experiment. The following proposition is of this character: It is in no way possible to completely reverse any process in which heat has been produced by friction."[17] Moreover, as Planck points out, the proposition touches all processes: "Since there exists in nature no process entirely free from friction or heat-conduction, all processes which actually take place in nature, if the second law be correct, are in reality irreversible."[18]

The significance of the Second Law lies not only in the assertion that a perpetual motion machine of the second kind is an impossibility, but also in stating unequivocally that all processes are irreversible.

The usefulness of the Second Law becomes more evident in the treatment of cyclic processes. No matter how complicated they are, they always give rise to a final state that is different from the initial state, however slight the difference—normally owing to some energy dissipation. The Second Law tells us that perfect cycles (reversible) are not possible in natural processes; once a process has taken place, the initial state is not recoverable.

Even the most reversible-looking processes turn out, when examined more carefully, to be irreversible. For example, Earth's rotation on its axis is slowing slightly and irreversibly due to tidal friction. It has been estimated that some 600 million years ago the daily rotation of Earth took approximately 21 hours. Frictional effects are slowing the planet an hour every 200 million years.[19] All processes occurring in the universe exhibit an irreversible behavior.

The Second Law applies to all natural processes—whether they are physical, chemical, biological, geological, or other processes. The strength of the law is based on the foundation that there is no known instance that contradicts the irreversibility of natural processes. Even a single violation would cause the entire edifice of the law to crumble—an all-or-nothing proposition. As Max Planck put it: "There is no third possibility."[20] On the other hand, every confirmation, even in seemingly remote regions, supports the whole structure.

Since the days of Kelvin and Clausius, more and more testimonial data have accumulated from all branches of natural sciences in support of full confirmation of the Second Law. It is hard not to be impressed by the simplicity and generality of the Laws of Thermodynamics. Einstein was no exception when he said:

A theory is the more impressive the greater the simplicity of its premises is, the more different kinds of things it relates, and the more extended is its area of applicability. Therefore the deep impression which classical thermodynamics made upon me. It is the only physical theory of universal content concerning which I am convinced that, within the framework of the applicability of its basic concepts, it will never be overthrown (for the special attention of those who are skeptics on principle).[21]

What Is That Quantity Called Entropy?

*From the point of view of philosophy of science the concep-
tion associated with entropy must I think be ranked as the
great contribution of the nineteenth century to scientific
thought.*[22]
— Sir Arthur S. Eddington

*There is no concept in the whole field of physics which is
more difficult to understand than is the concept of entropy, nor is there one which is more fundamental.*[23]
— Francis Weston Sears

The language of physicists is mathematics. Thus, it is not surprising that Clausius, a mathematical virtuoso, was the first to recognize that in Nature there must be a mathematical quantity that
distinguishes all irreversible processes from the reversible ones.
What is that mathematical quantity?

Both reversible and irreversible processes undergo changes.
What makes them different is the manner in which this happens.
Reversible processes, like a frictionless pendulum, return to their
exact initial state without having any additional effects upon their
surroundings. Irreversible processes, on the other hand, always
leave some effects in Nature. In essence, the mathematical quantity
in question must measure that effect.

Clausius discovered this quantity by first examining the works
of Joule and Carnot. Both had studied energy transformations,
which were begging for some precise mathematical quantification.
And that's exactly what Clausius did. He came up with a quantity
that measures the equivalent value of energy transformations, and
equated it to the thermal energy (Q) involved in the transformation divided by the absolute temperature (T). He denoted it by the
symbol S ($S = Q/T$), and called it the "transformation content" of a
body.

Reversible and irreversible processes can now be examined in terms of the quantity S. Since reversible processes return to their initial condition (state) after undergoing many changes, some transformations that occur must be positive in value and some negative, such that the sum of all the transformations becomes equal to zero. We can then say: In all reversible processes, there is no change in the value of S.

What happens to the value of S in irreversible processes? Clausius said that it always increases. Here is an example: In Joule's paddle-wheel experiment, an amount of work is performed by the descending weight, generating an amount of heat Q on the water at temperature T. The quantity S then increases by an amount equal to Q/T.

S increases because of the Second Law of Thermodynamics, which states that it is impossible to reverse the process—that is, to lower the temperature of the water back to its original value and at the same time raise the weight to its initial height—without leaving any other effect in the world. (If Joule's experiment were reversible, S would have remained constant, increasing by $+Q/T$ when the weight descended and the water warmed up, and decreasing by $-Q/T$ when the water cooled and the weight rose.) In fact, without the Second Law, we could even decrease S in the world. As noted, a ship could extract thermal energy from the ocean and use the energy to propel itself. Extraction of heat $-Q$ from the ocean at temperature T would change the quantity S by an amount $-Q/T$; that is, S would decrease. Again, this operation is impossible. In essence, the quantity S tells us which energy transformations are possible in Nature—those that increase the value of S.

Clausius did extensive theoretical work on the quantity S. In all the processes he examined, S increased. The significance of this thermodynamic quantity became obvious to him after more than a decade of theoretical development in association with Lord Kelvin. Brilliant scientists know when they stumble onto an important discovery. Clausius had found a quantity that always increases in all natural processes. The Second Law of Thermodynamics can now be set forth in a precise and general way. Instead of listing all the impossibilities under the Second Law, we can formulate the law concisely in terms of this quantity S. But first Clausius gave S a suitable name:

We now seek an appropriate name for S. . . . we would call S the transformation content of the body. However, I have felt it more suitable to take the names of important scientific quantities from the ancient languages in order that they may appear unchanged in all contemporary languages. Hence I propose that we call S the *entropy* of the body after the Greek word 'η τροπη, meaning "transformation." I have intentionally formed the word *entropy* to be as similar as possible to the word *energy*, since the two quantities that are given these names are so closely related in their physical significance that a certain likeness in their names has seemed appropriate.[24]

Every process in Nature, no matter how simple or how complicated, proceeds according to two Laws of Nature: the Law of Energy and the Law of Entropy. The first one says there exists in Nature a quantity called energy that remains constant. The second says there exists in Nature a quantity called entropy that always increases. Clausius summarized the long and arduous work of thermodynamicists by these extremely simple but far-reaching statements:

The energy of the universe is constant.
The entropy of the universe strives to attain a maximum value.[25]

Maxwell's Demon Attempts to Demolish Nature's Law of Entropy

The law that entropy always increases—the Second Law of Thermodynamics—holds, I think, the supreme position among laws of Nature. If someone points out to you that your pet theory of the universe is in disagreement with Maxwell's equations—then so much the worse for Maxwell's equations. . . . But if your theory is found to be against the Second Law of Thermodynamics, I can give you no hope; there is nothing to do but to collapse in deepest humiliation.[26]
—Sir Arthur S. Eddington

No demon or mortal has ever challenged the second law of thermodynamics and won.[27]
—George Musser, *Scientific American*

Ever since its discovery, the Second Law of Thermodynamics has been assaulted many times, on many fronts. Every time humans have expanded their stock of knowledge, they have made new attempts to destroy the Second Law and thus free themselves from its imposing constraints. Every challenge has come with a new twist.

One such attempt came from the renowned physicist James Clerk Maxwell, who in 1871 imagined an ingenious way to reverse the direction of entropy. The powerful weapon that was to crumble the Second Law was his newly formulated kinetic theory of gases.

In the mid-nineteenth century, Maxwell was the first to enunciate that temperature is a measure of the average kinetic energy of the particles that make up a system. On average, the molecules of a hot body move more rapidly and therefore are more energetic than those of a cold body.

When we insert a thermometer into a swimming pool, the reading indicates the average kinetic energy of all the trillions and trillions of water molecules that surround the thermometer. While at the microscopic level some molecules move faster and some move slower, at the macroscopic level of everyday experience we measure and feel the effect of the average kinetic energy of all the molecules.

Carnot's principle tells us that once all temperatures in a system have been flattened out, we can no longer extract work from that system. Since we now know that the energies of all the individual molecules are not equal, can we exploit this knowledge to build perpetual motion machines of the second kind?

Yes, said Maxwell, if we could find "a being whose faculties are so sharpened that he can follow every molecule in its course, and would be able to do what is at present impossible to us," which is to decrease entropy "without expenditure of work," in "contradiction to the second law of thermodynamics."[28] This "intelligent" being became known as Maxwell's demon. Maxwell thought that the demon, if given the inner workings of the kinetic theory of gases, should be capable of demolishing the Second Law.

The technique is simple. Imagine a system consisting of a gas at uniform temperature in an isolated enclosure. We put a partition down the middle of the enclosure and provide it with a small trapdoor. The demon's mission is to follow the movement of individual molecules and operate the trapdoor accordingly. When the demon sees a molecule with a speed greater than the average moving from left to right, or one with a speed less than the average moving from right to left, he opens and shuts the door instantaneously through a remote control mechanism. The idea is to send the more energetic particles to the right and the less energetic ones to the left. As time goes on, the right half of the container becomes continually hotter and the left half colder without expenditure of energy. When a certain temperature difference is built up, we can use it to drive a heat engine. Once we accomplish this, we obliterate the Second Law.

The paradox posed by Maxwell's demon bothered physicists for a few generations. Demons are not the kinds of creatures physicists like to deal with. But that is not the main reason. Maxwell's scheme strongly implied that someday humans can, through advanced technology, get around the Law of Entropy.

But there are many reasons why we cannot reverse the direction of entropy. The Second Law tells us that if we take account of everything, the entropy of an isolated system increases. In this case, the total system includes the container, the gas, the trapdoor, and the demon. The trapdoor is a mechanism that will dissipate some energy when it is opened and closed. Moreover, as the mechanism becomes smaller, the mechanism itself and its controls become subject to temperature fluctuations. The door may act erratically.

There is another flaw in Maxwell's scheme. How would the demon see the individual particles? As he peers into the isolated enclosure at uniform temperature, he cannot see anything because of the uniformity of radiation throughout. He needs some kind of flashlight to see the individual molecules. The introduction of the flashlight or any other source of radiant energy changes the whole picture. Now we have to account for this additional source of energy in our system. The energy of the flashlight is what provides the necessary "information" to operate the trapdoor. Otherwise, the demon cannot separate the high-velocity particles from the slow ones. As the Nobel laureate in physics Dennis Gabor put it, "We cannot get anything for nothing, not even an observation."[29] Thus if we account for every source of energy dissipation, we find that entropy still increases in our system. After all the trouble we went through, we have still not beaten the Second Law, not even with the help of "intelligent demons."

Physics Nobel laureate Richard Feynman demonstrated—in his celebrated book, *The Feynman Lectures on Physics*—the futility of trying to beat the Second Law at the microscopic level. Through detailed and careful analysis, he showed the impossibility of obtaining work from a system at uniform temperature (maximum entropy).[30]

These activities by physicists demonstrating that the Second Law cannot be violated have left some unpleasant impressions. Some disbelievers have taken the position that the Second Law is based on the inability of physicists and engineers to build a perpetual machine of the second kind. Max Planck found this position "untenable." He wrote:

> It would be absurd to assume that the validity of the second law depends in any way on the skill of the physicist or chemist in observing or experimenting. The gist of the second law has nothing to do with experiment; the law asserts briefly that *there exists in nature a quantity* [entropy] *which changes always in the same sense in all natural processes.* The proposition stated in this general form

may be correct or incorrect; but whichever it may be, it will remain so, irrespective of whether thinking and measuring beings exist on the earth or not. . . . The limitations to the law, if any, must lie in the same province as its essential idea, in the observed Nature, and not in the Observer. That man's experience is called upon in the deduction of the law is of no consequence; for that is, in fact, our only way of arriving at a knowledge of natural law. But the law once discovered must receive recognition of its independence, at least in so far as Natural Law can be said to exist independent of Mind. Whoever denies this must deny the possibility of natural science.[31]

The Second Law is not subject to the control of our technologies—no matter how advanced they become. While it is possible to counteract the effects of other laws of Nature—like gravity—through technology, it is impossible to reverse the direction of entropy increases by any means whatsoever. The Second Law is in absolute command.

Entropy as "Time's Arrow"

In the time of your life, live.[32]
—William Saroyan

Humans have devoted considerable time to developing and understanding the elusive physical concept called "time." In physics, a major problem in defining time stems from the fact that time is a basic concept, which means it cannot be expressed in terms of other concepts. Every discipline must start with some concepts. Physics starts with mass, length, and time. These concepts are taught and learned by talking about them, by showing examples, and by demonstrating objects that measure them—scales for mass, rods for length, and clocks for time.

Before physics became a field of study, the early Greek philosophers had attempted to develop the concept of time. As usually happens in philosophical discussions, however, they started many controversies. One argument concerned "becoming" or "change" versus "being" or "permanence." One school of thought put the emphasis on change, while the other stressed the static being. Heraclitus was among the philosophers who were convinced that only flux, change, and becoming are real; that permanence and constancy are merely apparent. "You cannot step twice into the same river; for fresh waters are ever flowing in upon you."

Plato attributes to Heraclitus the views that "nothing really is, but all things are becoming" and that "all things flow and nothing stands." Everything that appears static to us embodies, when properly understood, continuous movement. Should this activity stop, the universe would collapse to nothingness.

Philosophers who disagreed with Heraclitus' line of thought included Zeno of Elea and Parmenides. They believed that only the permanent and the lasting are real. All flux, time, change, and motion are unreal. Parmenides argued that in accepting change, Heraclitus had presupposed that something can both be and not be. According to Parmenides, if anything is, it is now, all at once.

The early Greek philosophers had a solution to the problem of time. If time should require both change and permanence, they would make one aspect real and the other apparent. For Parmenides and Zeno, constancy was real and change was unreal, while for Heraclitus it was the other way around. Thus the problem of time had been settled at the expense of dividing the universe into two parts: the constant and the changing.

A century later, Aristotle examined time more analytically. He was greatly concerned with the properties of time—especially its measurability—rather than in its status as real or apparent. He introduced two additional properties related to time. When water changes from cold to hot, the process can be understood quantitatively by its rate of heating. In addition, the water that was cold "before" became hot "after." Thus time, which cannot exist without change, has both an arrow-like as well as a durational character. The unit of time takes the form of a vector whose direction is fixed and whose length indicates lapse of time. Aristotle then arrived at his definition of time as "the number of motion in respect to 'before' and 'after.'"[33]

Today, physical laws encompass all the properties of time that the Greek philosophers conceptualized and developed. Of course, the terminology is different. Physics uses Laws of Nature expressed in mathematical language to describe the properties and behavior of time.

Modern philosopher Hans Reichenbach was convinced that physics is the right place to look for the resolution of time. He wrote: "There is no other way to solve the problem of time than the way through physics. More than any other science, physics has been concerned with the nature of time. If time is objective the physicist must have discovered that fact."[34]

As it turns out, physics, like Greek philosophy, also divides the universe into two parts: processes that are reversible, such as purely mechanical systems, and processes that are irreversible, which cover all natural processes. Our world presents itself with both processes: the uniform cyclic changes of reversible processes, which allow us to build accurate clocks, and the progressive, evolving changes of natural processes, with no exact repetition of states. The nature of our time concept rests ultimately on the existence of these two kinds of phenomena: one that gives uniformity and recurrence for the measure of time, and one that makes time the associate of flux, change, and novelty. While we measure the passage of time through cyclical repetitive events, we experience time through irreversible natural processes.

Modern accurate timekeeping started with Galileo's discovery of a periodic process that can be counted accurately—the swinging pendulum. In his later life, Galileo contemplated applying the pendulum to clocks by mechanically recording the number of oscillations.

In 1656, Galileo's desire was fulfilled by Christian Huygens. His pendulum clock, the first reliable piece of machinery to measure time with an accuracy of 10 seconds a day, inaugurated the era of high-precision timekeepers. Since then, timekeeping devices have become increasingly accurate, but the principle of accurate measurement of time has remained the same. We rely on the existence of cyclic processes that are uniform, as free of friction and dissipative forces as possible, involving little change in entropy. A perfect pendulum with no friction whatsoever would keep its period of oscillation constant. From such an ideal device, we get the feeling of constancy, permanence, and perpetuity as envisioned by some Greek philosophers.

The invention of mechanical clocks that could tick away for years on end greatly influenced the thinking of scientists, who began to believe in the uniformity and continuity of time. Sir Isaac Newton was convinced of the existence of an absolute time, which flows at a uniform rate regardless of anything that goes on in the world. He wrote in *Principia*: "Absolute, true, and mathematical time, of itself, and from its own nature, flows equably without relation to anything external."[35]

Until the beginning of the twentieth century, the concept of absolute time remained dominant—although Leibnitz and others had argued against it. This belief in a Newtonian absolute time was not confined to scientists alone. In industrial nations, life became regulated by the amazing clock, particularly after the mass production of inexpensive watches.

Starting in 1905, Einstein made major modifications to Newtonian physics. His special theory of relativity changed the idea of time from absolute to relativistic. In the process, however, time became a more abstract and complicated concept. Einstein declared that space and time are not independent entities, but should be viewed as two sides of the same coin, which we may call space-time. What glues the two sides together is the speed of light. The three-dimensional space, composed of depth (x), width (y), and height (z), became bonded with time (t) and the speed of light (c) through a four-dimensional space-time entity (s). The formula that connects these elements had been established by the Dutch physicist H. A. Lorentz: $s^2 = x^2 + y^2 + z^2 - c^2t^2$.

Einstein's mathematics teacher Hermann Minkowski was fascinated by the Lorentz transformation formula. In 1908, he enthusiastically announced the undissolvable marriage between space and time: "Henceforth space by itself, and time by itself, are doomed to fade away into mere shadows, and only a kind of union of the two will preserve an independent reality."[36] He argued that "nobody has ever noticed a place except at a time, or a time except at a place."[37]

Einstein was impressed by his teacher's eloquent arguments. "It appears, therefore, more natural," Einstein wrote, "to think of physical reality as a four-dimensional existence, instead of, as hitherto, the *evolution* of a three-dimensional existence."[38]

All those sophisticated, precise timekeeping devices and procedures are fine and useful, but Aristotle reminded us 24 centuries ago that time has an arrow-like property. We need, then, a mechanism—a theory, a Law of Nature—that can tell us the direction that time flows. The crucial question remains: Can classical or relativistic mechanics distinguish for us the past from the future? The answer is no. But why?

A good way of demonstrating the reason behind it is to watch a movie showing a purely mechanical system in action. Imagine a film showing a pendulum swinging freely from one side to the other, or Earth going around the Sun clockwise. Now, if we reverse the movie, we will watch a pendulum swinging back and forth, or Earth going around the Sun counter-clockwise. The movie, played backward, will appear normal. We will not detect anything blatantly wrong because the processes are reversible.

Now suppose we watch a film of a lit cigarette burning on an ashtray until it becomes ashes. If we play the film backward, we will see ashes and smoke turning into a cigarette, definitely contrary to everyday experience. Similarly, if we see a scrambled egg unscramble, we will immediately recognize that something is abnormal with the movie because the process is irreversible.

These movie demonstrations indicate that some processes are reversible, while others are irreversible. In fact, the word "reversible" means with respect to time. In Newtonian physics, as well as in relativistic mechanics, objects or particles can—in theory—move forward or backward in time. For this reason, we do not detect anomalies when we reverse a film showing reversible systems.

It should be emphasized that even the reversible systems are in actuality irreversible. All pendulums, due to frictional losses, gradually slow down and eventually stop. If we had a very long film and recorded the swings of the pendulum until they came to a halt and then played the film backward, we would know something is wrong. We would see a pendulum at rest spontaneously move by itself and move faster and faster with greater and greater swings—a happening never witnessed in the natural world.

Are mechanical systems the only ones that are time-reversible? No, the laws of electricity and magnetism are also time-reversible. How about the laws of nuclear and particle interactions? "With one trivial-seeming exception," the decay of K-mesons, "the laws governing the *micro*world are reversible in time," notes cosmologist Martin Rees.[39] (The rare and unstable particles called K-mesons occasionally—in 0.2 percent of cases—decay in a manner that violates time symmetry; James Cronin and Val L. Fitch received the Nobel prize for this discovery.)

Time-reversible processes are useful for making clocks and keeping time; nevertheless, they cannot tell us the direction of time's flow. We have to look elsewhere for that. How about the Laws of Thermodynamics? The First Law is not much help because it maintains total neutrality (symmetry) as far as time is concerned. When we say energy is conserved, we mean with respect to time. If we ask the question: What was the energy of the universe yesterday? The answer is: the same as today and tomorrow—the same as always. Thus the First Law tells us nothing about the flow of time.

We need then a quantity that varies with time. Such a quantity must be unidirectional, since time flows only one way. It must also be consistent with everyday experience. Immediately, entropy becomes an excellent candidate for telling us the direction of time, since the Second Law of Thermodynamics unequivocally states that in all natural, irreversible processes, entropy always increases.

Thus, if we ask: What was the entropy of the universe yesterday compared with today's and tomorrow's? The answer is: yesterday, entropy was less than today; tomorrow, it will be greater than today. As time flows from past to future, entropy always increases in the same direction. Sir Arthur Eddington vividly expressed the situation when he said that entropy increase gives us the direction of "time's arrow."

Boltzmann's Entropy Relation

When we add cream to coffee we see everything blending. But it would be quite a spectacular event if we saw the cream and coffee suddenly unmix. Why does this kind of thing never happen? Liquids mix, gases expand, but the reverse never occurs. Can the explanation be found in the Law of Increasing Entropy?

That is what Ludwig Boltzmann tried to do toward the end of the nineteenth century. In the process, he started such a wrangling that to this day we can still hear some reverberations. It was not the principle of entropy that got Boltzmann into trouble; it was his new brand of physics and mathematics, pushing physics beyond what could be accepted at the time.

The attack on Boltzmann's theoretical work came from many directions. The mathematicians J. Loschmidt and E. Zermelo criticized the mathematical and philosophical foundations of his work. Powerful physicists like Ernst Mach and Wilhelm Ostwald criticized the underlying model of atoms and molecules. In a series of articles, Mach "attacked the use of atomic models in physics, and asserted that the basic purpose of science is to achieve 'economy of thought' in describing natural phenomena, not to explain them in terms of hypothetical concepts such as atoms or ether."[40]

In short, some serious charges were leveled against the theoretician who once had been a skillful experimentalist but had to leave a promising career in experimental physics because of growing nearsightedness. In 1898, keenly aware of the growing animosity toward his work, Boltzmann wrote: "I am conscious of being only an individual struggling weakly against the stream of time."[41]

As the leader of the atomist school, Boltzmann frequently had to engage in heated debates with the members of the school of Energetics (Ostwald, Mach, P. Duhem, and others) who had proposed developing thermodynamics without any reference to atomic theories. The constant attacks on Boltzmann finally took their toll. In 1906, suffering from acute headaches and fits of severe depression, Boltzmann committed suicide. It was not long, however, especially after Jean Baptiste Perrin's experimental work on Brownian motion, before the scientific community made a complete turnabout and accepted the idea that atoms exist. By 1909, even Ostwald had become a convert to Boltzmann's point of view. Ostwald also had some kind words for Boltzmann, calling him "the man who excelled all of us in acumen and clarity in his science."[42]

What made Boltzmann's approach so unacceptable? Three serious accusations were brought up against him. First, he used an atomic model of matter not fully accepted as scientific gospel. His second sin was the introduction of mechanical models into thermodynamics. Boltzmann's antagonists argued that mechanical systems and equations are all reversible, yet thermodynamic phenomena most certainly are not. How can one start with a mechanical reversible model and end up explaining an irreversible thermodynamic system?

It can indeed be done, said Boltzmann. To prove his point, he used probability theory to show how one can start with reversible phenomena at the microscopic (atomic) level and end up with irreversible phenomena at the macroscopic level. And this was the third sin he committed. Since the days when Newton discovered the law of gravity, all physics was based on certainties and not on probabilities.

Boltzmann, on the other hand, pleaded with his opposition not to pass a hasty judgment. He was convinced that the attacks on his theory were "merely based on a misunderstanding," and held that if physicists paid careful attention to what he was saying they would find the explanation to the irreversibility paradox.[43] Eventually, physicists did exactly that.

Suppose we take a box and put a barrier in the middle. We fill one side with neon gas ("black" molecules), and the other with argon ("white" molecules). We all know what happens when we pull the barrier out. They mix. Once the barrier is removed, the particles are free to roam around in a greater volume. Gradually a white molecule goes toward a black one and a black one toward a white. Molecules collide and bump each other. If we wait long enough, we get a homogeneous mixture. This is an irreversible process, which leads to an increase in entropy.

We have here an example of what was once referred to as the "irreversibility paradox." We have an irreversible process, yet each separate event in the process is reversible. Molecules in a gas are in constant motion, sometimes colliding with each other. Every time we have a collision between two molecules, they go off in certain directions. If we take a moving picture of a particular collision and then show the picture in reverse, nobody would find anything wrong with it; one type of collision is just as likely as another.

Now comes the apparent paradox: Every collision obeys the reversible laws of mechanics and yet, in totality, the mixing of the two gases is an irreversible phenomenon. How can we have irreversibility based on reversible situations? Isn't that a bit odd? Yes, indeed. "But we also see the *reason* now" said Richard Feynman. "We started with an arrangement which is, in some sense, *ordered*. Due to the chaos of the collisions, it becomes disordered. *It is the change from an ordered arrangement to a disordered arrangement which is the source of the irreversibility.*"[44]

So if we watch a movie that shows individual collisions and see the picture in reverse, we will not detect anything wrong. But if we watch a picture of all the collisions, and then watch the movie backward and see the gases unmix, we will not only smile, because it violates our everyday experience, but we may also exclaim: "It's against the laws of physics! The Law of Increasing Disorder!"

So far, Boltzmann has been exonerated on two counts: His atomic and mechanical models have both been cleared of wrongdoing. How about the introduction of probability into physics? Boltzmann pointed out that if the number of particles is sufficiently large, mechanical processes are irreversible. He explained it this way: If we shake together 1,000 black marbles with 1,000 white marbles, we obtain a random mixture. Further shaking does not reverse the process.

Boltzmann's example, of course, is purely an analogy, since there is no shaking going on in the mixing process of the two gases. What he was trying to show us was the effectiveness of the "law of large numbers." As the number of molecules in a gas gets larger and larger, the probability of witnessing the re-separation of neon and argon molecules becomes smaller and smaller. And when the number of molecules becomes astronomically large, as in the real world, the time we have to wait for the gases to unmix surpasses many times over the estimated lifetime of our universe. There is no reason, then, to hold our breath; we will never see gases unmix, or liquids, for that matter.

In statistical mechanics, physicists have given a special meaning to "disorder." Physicists measure disorder (Ω) by the number of ways (microsituations) a given thermodynamic situation can be realized; the greater the number of microscopic states (called complexions) a system can assume, the greater its disorder. (The more ways a system can be arranged, the greater its disorder.) Suppose we had only one molecule or one marble and one box to put it in; we have only one way of arranging the system. If we divide the box into two compartments, we have two ways of arranging the system; we can put the marble in the left compartment or in the right one. If we had two identical marbles, we can arrange the system four ways: both marbles in the left compartment, both marbles in the right compartment, marble 1 in the left compartment and marble 2 in the right one, and vice versa. As the number of molecules or marbles increase, the number of ways of arranging the system increases accordingly. Boltzmann enunciated that the natural logarithm (ln) of that number of ways is proportional to the entropy S of the thermodynamic system. However, he did not discover the physical constant of proportionality (k) that bears his name. It was introduced by Max Planck. Here is the entropy expression that caused Boltzmann so much agony:

$$S = k \ln \Omega \qquad \text{Boltzmann's entropy relation}$$

It has been carved above his name on his tombstone in the Central Cemetery in Vienna.

Although today physics has attained a state of stunning complexity, simplicity remains its major goal. Physicists love to come up with simple equations that describe the Laws of Nature. In Feynman's words, "You can recognize truth by its beauty and simplicity."[45] This is one reason Boltzmann's entropy relation received so much attention by physicists. It passed the aesthetic test. But that is not enough. Feynman describes the real test a new law must pass to gain acceptance as a scientific truth:

> First we guess it. Then we compute the consequences of the guess to see what would be implied if this law that we guessed is right. Then we compare the result of the computation to nature, with experiment or experience, compare it directly with observation, to see if it works. If it disagrees with experiment it is wrong. In that simple statement is the key to science.[46]

Boltzmann's entropy relation has survived this long because it works. Its widespread use stems not only from the simplicity of the equation but also from the agreement between its theoretical predictions and actual experimental observations. Boltzmann's relation has made significant contributions to the advance of science, from chemistry to information theory. One major contribution, Planck's discovery of the quantum laws of radiation, owed much to Boltzmann's statistical theory of entropy.

Chapter   3

Nature's Laws in Action

Man continually engages in attempts to create order, but only at the expense of greater disorder in the surroundings.[1]
—G. Tyler Miller, Jr.

The Relentless Increase of Entropy

It has been said that science begins with observation. Science advances, however, not merely by accumulating mountains of scientific facts but by asking why phenomena occur as observed. Scientists then attempt to discover laws that govern such phenomena.

Chemists had always wondered why some reactions or changes occur spontaneously while others are either forbidden or can only occur through some inducement—that is, through the addition of some form of energy into the system. For instance, why do cold hydrogen atoms unite, and hot hydrogen molecules disintegrate, and phosphorous burst into flame when exposed to air? It is because of "nature's determination to increase the total entropy at all costs," remarks Sir George Porter, a Nobel laureate in chemistry.[2]

Before the advent of thermodynamics, chemistry had no general laws that could explain in a coherent way the behavior of all chemical phenomena. Thermodynamics provided a theoretical framework through which chemists could explain and predict the behavior of chemical reactions and changes.

Chemists found out that all chemical reactions proceed in accordance with the First and Second Laws of Thermodynamics. The First Law, when applied to chemistry, states that in all chemical processes the total energy of the system—chemical reactants and products—and its environment remains constant. Although this is useful for bookkeeping purposes, it does not explain why some chemical changes occur spontaneously while others do not.

The question remains: Why do things change in the direction they do? After years of theoretical and experimental work, chemists found that the Second Law is responsible for the way chemical reactions and changes occur. Nature's tendency to increase entropy is the Supreme Law that governs all chemical behavior.

Chemists calculated the overall change in entropy of a variety of chemical reactions that occur spontaneously. In all cases, they found that the total entropy (system and surroundings) always increases, in accordance with the Second Law. Suppose we want to know whether gaseous hydrogen and oxygen would combine to form steam at 100 degrees Celsius and at constant pressure. Calculations show that the entropy change for the system is approximately equal to −11 units of entropy (calorie per degree Kelvin), while the entropy change for the surroundings is approximately +150 units of entropy.[3] Since the net entropy change is positive, according to the Second Law the reaction should occur spontaneously. In fact, it does. Moreover, whenever the net change in entropy is negative, the Second Law states that the process cannot happen spontaneously, and it never does.

Whether we are dealing with chemical reactions that occur naturally or are produced by humans, the result is the same—an increase in entropy of the thermodynamic system. This means that the more chemicals we produce and use, the more we increase the disorder around us. Today's environmental problems are indeed consequences of the Second Law.

The Second Law tells us there is in Nature a constant tendency for order—low entropy—to turn to disorder. In everyday life, we hear and use the words order and disorder. As it turns out, physicists have given these words meanings not much different from those we intuitively use in our daily activities.

For example, a collection of objects arranged in a definite pattern, like a neat pile of blocks sitting on a table, is considered an ordered arrangement. The moment a child begins to play with this neat stack of blocks, the pieces become distributed all over the table in a disordered, haphazard, or random manner. And once the ordered arrangement has been shattered, we do not expect the blocks to return spontaneously to their original, ordered state. This type of disorder is referred to as positional disorder. Whenever objects are lined up or stacked in a small volume, there is a low degree of positional disorder (more order). And whenever the objects are randomly distributed over a larger volume, there is a greater degree of positional disorder (less order).

In general, particles and molecules do not like to be confined. Given the choice, they like to roam around freely. Suppose, for example, we take all the air from a hermetically sealed room, put all the gas molecules into a small bottle, and insert a stopper. Now suppose we remove the stopper. What will happen? The room will fill with air, as the molecules choose not to stay cooped up in the small bottle but instead to wander freely within the larger volume.

Boltzmann's equation explains this phenomenon. The molecules of the air can distribute themselves in more ways within the larger space of the room than in the smaller volume of the bottle. The more ways a given state can be realized, the greater its statistical probability. And because the logarithm of the statistical probability of an aggregate is a measure of the aggregate's entropy, Boltzmann's equation says that the entropy of the air is greater in the room than in the bottle. This is in conformity with the Law of Entropy—the general tendency for disorder to increase.

Societies apply the same overall principle to maintain order. We build prisons to confine into a small area the movement of a certain group of people who, in society's judgement, generate unacceptable amounts of disorder of one form or another.

Tiny molecules in a gas increase entropy through expansion, which makes their distribution uniform. On the other hand, massive gravitating bodies—due to their immense attractive gravitational forces—increase entropy through gravitational compaction. In these processes, the biggest increase in entropy is achieved when a system of gravitating bodies collapses into a black hole. The equation for calculating the entropy of a black hole is due to the work of Jacob Bekenstein and Stephen Hawking.[4] The entropy of a black hole turns out to be proportional to the square of its mass.

Thus, the entropy per unit mass of a black hole is simply proportional to its mass, and gets larger when the black hole gets larger. "Hence," writes mathematician and astrophysicist Roger Penrose of Oxford University, "for a given amount of mass—or equivalently, by Einstein's $E = mc^2$, for a given amount of *energy*—the greatest entropy is achieved when the material has all collapsed into a black hole! Moreover, two black holes gain (enormously) in entropy when they mutually swallow one another up to produce a single united black hole!"[5] Here on Earth, large corporate entities increase their entropy (and the entropy of their constituents) when they mutually swallow one another up—in friendly or hostile takeovers—to form a single huge conglomerate.

Another way to increase entropy in the system is to leave the volume unchanged but change the number of ways that molecules can move around in the same volume. Solids are lower in entropy than liquids, which in turn are lower in entropy than gases. Why? Because the molecules in solids have the least amount of individual freedom to move around, while the molecules of gases have the greatest.

Entropy also increases when the number of molecules increases within a given volume (space).[6] If we apply this situation to human beings, we see that the world's overall entropy is increasing as global population is in a state of explosion. Moreover, thanks to technological advances, each individual's capacity to create disorder is also growing. Consequently, it is becoming harder to maintain order both locally and worldwide.

Keeping the number of molecules the same but pumping energy into the system also causes the entropy of the system to increase. Why? Because the molecular velocities increase due to the addition of energy. Our world is becoming ever more highly entropic as humans develop technologies that can move materials, goods, chemicals, weaponry, people, machines, and messages faster and faster.

Human beings can also be classified into low-entropic and high-entropic people. Psychologists call high-velocity, high-energy achievers Type A. These individuals attempt to accomplish too much, are involved in numerous activities, and cram more and more things into less and less time. They live in the fast lane. Eventually, the fast-tempo, high-entropic life-style begins to take its toll from the human body. The wear and tear may manifest itself in the form of ulcers, heart attacks, nervous breakdowns, or other such disorders. Friends, psychologists, and doctors alike give the same advice. They tell Type A people to slow down (lower their entropy) and become Type B.

One way to add energy to a thermodynamic system is to add heat. This makes molecules move faster, thus increasing the system's entropy. In thermodynamics, a cooler system (or environment) is more orderly than a heated one. We consider a heated situation chaotic and dangerous. We use the phrase "cool heads prevailed" whenever a potentially disastrous situation is avoided by the intervention of calm, low-entropic people. We also use such expressions as "we are putting on the heat" or "we are turning up the heat" whenever we want to convey the idea that we are exerting pressure on someone.

Heat is a measure of the energy and internal disorder of a system, as exemplified by this witty *Wall Street Journal* definition, "Heated economy: thermostate of the union."[7] When a system, whatever it may be, is overheated and becomes out of control, we use the verb "freeze" in an attempt to put some order into the system. For example, when the U.S. budget became too disorderly, with the federal deficit increasing uncontrollably, Senator Fritz Hollings of South Carolina called out to "freeze the budget" in an effort to restore some order to the U.S. government's chaotic financial activities.

The disintegration or decomposition of molecules is yet another way through which the entropy of a system increases. In Nature, there is a tendency for molecules to break up and disintegrate. As it turns out, the same overall situation prevails in human beings. There is a general inclination for families to break apart. Members of the family, parents, and society make constant efforts to keep the nucleus from disintegrating. Laws are passed to hinder the dissolution of marriages. Some religions go so far as prohibiting divorce all together. When a society's divorce rate increases, its overall entropy grows, as the more frequent family breakups generate a greater level of social disorder.

One question often raised is, If it is true that there exists in Nature a general tendency for order to turn into disorder, then how could Nature have produced such highly organized entities as molecules, crystals, and organisms? The processes that created these entities, are they in violation of the Second Law of Thermodynamics? Not at all.

The Law of Entropy does not prohibit the creation of highly organized, orderly things like diamonds and crystals. Nature does some work (expends some energy) while creating the crystal, and in so doing generates more disorder in the environment than the order within the crystal. After the formation of the crystal, the entropy of the thermodynamic system is greater, in conformity to the Second Law.

Like Nature, humans can also fabricate crystals and diamonds. Today's manufactured diamonds are constructed so well that they look very natural. The point to remember, however, is that in the process we generate more disorder in our environment than the order we create in the diamond. Similarly, when humans produce a copper sheet from some copper ore, the decrease in the disorder (entropy) of the ore is more than compensated for by the entropy increase in the rest of the environment. In the case of gold, of the

7.235 billion tons of gold ore mined in the United States in 1998, only 0.00033 percent ended up as gold, the remainder ended up as waste in the environment, much of it toxic.[8] The Second Law says we can create order from disorder as long as we are willing to pay the price of increasing the overall entropy of the world.

Whether humans or Nature creates order out of disorder, or disorder out of order, some work must be done in the process and some energy expended, thus increasing the overall entropy of the universe. Modern human beings are rapidly increasing the overall disorder of their thermodynamic system while believing, all along, that they are creating order in an otherwise disorderly environment.

Although order can be created out of disorder and vice versa, it takes much less effort and time to generate disorder from order than the other way around. It takes more effort to put order into an apartment or house than it does to make a mess out of it. It requires months and sometimes years, and a considerable amount of energy-matter transformations, to build a home, a car, a computer, an airplane, a chemical plant, or a nuclear reactor. Yet one fire, one accident, or one natural mishap is enough to convert them quickly to rubble.

From a Clockwork Universe to the Heat Death of the Universe

"What is there? Everything," said the noted philosopher Willard Van Orman Quine.[9] Scientists call "everything there is" the universe. Through observation and experimentation, they try to discover laws that explain and predict what goes on in the universe. Most laws, however, apply to a small portion of the observed phenomena; the laws of electricity explain only the behavior of electrical phenomena. Other laws are more general. The First and Second Laws of Thermodynamics are so general that they apply to everything known.

Some laws tend to evolve into something larger, a paradigm or model into which all future observations, theories, and laws must fit. Before the discovery of the Laws of Thermodynamics, the mechanistic worldview prevailed as a paradigm emanating from the work of Galileo, Kepler, and especially Newton.

In the late seventeenth century, Sir Isaac Newton discovered the universal law of gravitation. He started with the modest assumption that the falling of everyday objects, like apples and rocks, has something in common with the movement of heavenly objects like planets and stars. He then developed a single mathematical formula—beautiful for its simplicity—that describes the motions of all bodies in the universe.

According to Newton's law of gravitation, every particle of matter in the universe interacts with every other particle through the force of gravity. Newton's great achievement was the discovery of the physical characteristics of the force of gravity: the force between every pair of objects is attractive, proportional to the masses of both objects and inversely proportional to the square of the distance between them. If the mass of a body is doubled, its gravitational pull on the other body doubles. If the distance between two bodies is doubled, the gravitational force between them reduces by a factor of four.

Even before Newton's birth, Johannes Kepler had already announced the laws of planetary motion. He had found them empirically after 30 years of tedious calculations using Tycho Brahe's extensive and accurate planetary observations. He did not know why planets move the way they do, except he thought the Sun might have something to do with it. Newton, on the other hand, using simple mathematics, deduced from Kepler's lifelong work the reason planets move the way they do. He put science on a theoretical foundation.

Newton's discovery inspired confidence, helping us realize that the human mind is capable of unraveling Nature's mysteries and understanding regularities in the apparently chaotic world in which we live. The success of the universal gravitational law in describing the solar system led to the vision of a "clockwork" universe, the mechanics of which could be well understood. The universe was no longer looked upon as a chaotic place, but a place where "law and order" prevailed. Everything moved according to a master plan; the motion of every object was completely determined. The French astronomer and mathematician Pierre Laplace put it this way:

> An intelligence which at a given instant knew all the forces acting in nature and the positions of every object in the universe—if

endowed with a brain sufficiently vast to make all the necessary calculations—could describe with a single formula the motions of the largest astronomical bodies and those of the lightest atoms. To such an intelligence, nothing would be uncertain; the future, like the past, would be an open book.[10]

Such a hypothetical intelligent entity that could know the past as well as the future with certainty became known as "Laplace's demon." To our knowledge, no such demon exists; still, the gravitational law at least suggests that in principle all of Nature may work in this deterministic fashion.

The Law of Entropy drastically changed our view about Nature. The gravitational force or gravitational energy, nondissipative in character, implies that once all objects in the universe have been set in motion, they will go on forever without any degradation. The First Law did not alter this picture. The fatal blow to the mechanistic view of Nature came from the Second Law. The concepts of entropy, irreversibility of natural processes, and dissipation of energy-matter in the Second Law were too much to take at one time. The mechanistic worldview could not absorb that big of a punch and survive. It started reluctantly and slowly to give way to the entropic worldview.

Lord Kelvin was the first to apply the Second Law to Nature as a whole. In a historic paper, Kelvin pointed out that when heat is produced by an irreversible process, there is a dissipation of energy, and a full restoration of the initial condition is impossible. He continued by giving other examples of energy dissipation, such as when heat is diffused by conduction. Kelvin concluded his paper with this bleak outlook:

> Within a finite period of time past, the earth must have been, and within a finite period of time to come the earth must again be, unfit for the habitation of man as at present constituted, unless operations have been, or are to be performed, which are impossible under the laws to which the known operations going on at present in the material world are subject.[11]

Throughout history, we humans had convinced ourselves that we were placed at the center of the universe because we were special. Then Copernicus, Galileo, and Kepler came along and stated that Earth moves around the Sun, and not vice versa. This statement of fact was a major disappointment. Our place of habitation was no longer the focal point of the universe. Then Kelvin came along and stated that Nature's energy is gradually dissipating, and a day will come when Earth will be unfit for habitation. This unpleasant state of affairs heightened our disillusionment.

A few years later, Helmholtz gave the knockout punch to our long-term aspirations. He said that the universe is "running down" and will finally reach maximum entropy, at which point all available energy will be used up. All temperature differences will be flattened out, and the universe will reach a state of absolute uniformity throughout, where no energy transformation can take place. When and if such a state is ever reached, although the total amount of energy in the universe will be the same as always, all natural processes would have to cease. The end point toward which we are headed—as envisioned by Helmholtz—is called the "heat death" of the universe.

Thermodynamics and Cosmology

Everything that happens in the universe consists of the same basic stuff, "mass-energy," transfigured in space and time from one form into another.[12]
—Craig J. Hogan

The Laws of Thermodynamics play an important role in the world of astronomy. Astrophysicists view the cosmos as being composed of many forms of energy that are continuously undergoing transformations. The principle of entropy strongly suggests that each form of energy has associated with it a characteristic "quality" measured by its entropic content; the higher the quality of an energy type, the lower its entropic content.

Gravitational energy is the highest quality energy form in the universe, followed by nuclear energy, internal heat of stars, sunlight, chemical energy, terrestrial heat waste, and finally cosmic background radiation, the lowest quality.[13] In fact, gravity, which is nondissipative, carries no entropy. Hence, a hydroelectric power station can convert the gravitational energy of a waterfall into electricity at efficiencies nearing 100 percent, which no nuclear power plant can approach.

The universe evolves by undergoing energy transformations. In general, higher forms (quality) of energy are degraded into lower forms in energy transformations—generating entropy in the process. If there were only gravitational energy, the universe would be like the perpetual clockwork machine envisioned by seventeenth-century mechanists. Entropy would never increase, and the universe would never age. At the other extreme, if in due time everything ends up as background radiation, the universe will then have reached its heat death—maximum entropy—envisioned by nineteenth-century thermodynamicists. Having reached final equilibrium, the universe will then be "just black, empty, expanding space, for the whole of future eternity."[14]

In recent years, two cosmological theories have emerged about the origin and evolution of the universe: the big bang and the steady-state (continuous creation) theories. The big bang theory was first conceived by a Belgian priest and astrophysicist, Georges Lemaître. In 1927, he suggested that all matter was once squeezed into a single object, which he called the "primeval atom."[15] This became unstable and exploded, driving the material out to form the galaxies. The Russian-born American physicist George Gamow was in agreement with this scenario. Together with his students Ralph Alpher and Robert Herman, he proposed that the universe began very hot with a big explosion, which is now called the "big bang." This is a popular model. The highly concentrated inferno exploded with such a force that the universe has been expanding ever since.

In 1948, three British astronomers—Hermann Bondi, Thomas Gold, and Fred Hoyle—came up with a radically different cosmological view, known as the steady-state model. Their proposal was based on a provocative principle: that on a very large scale, the universe as a whole remains approximately the same from epoch to epoch. While some galaxies and stars irreversibly evolve, decay, and burn out, other galaxies are born to replace them. They argued that a continuous "creation of matter out of vacuum" is taking place in the universe, which replenishes at a steady state the dissipation of energy and dispersion of matter.[16] Their cosmological model shatters the First Law of Thermodynamics as a principle valid for the universe as a whole.

Both cosmological models conform to the overwhelming physical evidence that the universe is expanding. As early as 1929, Edwin Hubble confirmed that galaxies are moving away from us at enormous speeds. While looking at distant galaxies, he observed a phenomenon called the Doppler effect. Whenever a light source moves away from the observer, the detected electromagnetic frequency shows a shift toward the red end (lower frequency) of the light spectrum. The magnitude of the frequency shift reflects how fast the observer and source are receding from each other.

Hubble noticed that light from distant, faint galaxies is systematically more red than that from closer, brighter galaxies. This was an indication that distant galaxies are receding from us. Hubble's discovery was the first hard evidence that the universe is dynamic and expanding, and gave credence to the theory that the universe began with a violent explosion. It also overturned centuries of misconception about the universe, which had been thought to be a static place. Since then, Hubble's findings have been confirmed many times.

Can a universe whose distant galaxies are hurtling away at enormous speeds be pulled back? The only energy we know that could accomplish such a Herculean task is gravitational energy. But no matter how astronomers have counted, they have not been able to find enough mass to pull the galaxies together and to close (collapse) the universe. The gravitational glue that is needed to keep the universe from expanding to its eventual heat death has yet to be found.

However, there may be massive amounts of invisible or "dark" matter scattered throughout space, whose gravitational pull could eventually collapse the universe. For this to be possible, there would have to be many times as much dark matter permeating space as can be directly identified through telescopes.[17] There is indirect evidence that a significant amount of dark matter is present, but whether there is enough of it to collapse the universe—a process called the "big crunch"—is an open question.

Based on today's evidence, the universe will expand indefinitely. However, could it be that as the universe expands, some galaxies evolve, decay, and burn out, while others are born in such a way that the global aspect of the universe remains the same from epoch to epoch? The steady-state model, referred to as the perfect cosmological principle, states that the universe always looks the same.

The steady-state model so intrigued astronomers that in the early 1960s they began a systematic survey to test its validity. The experiments consisted of counting the number of radio sources as a function of epoch, back through time. The results showed that there were more radio sources in the distant past than there are now. This finding was devastating to steady-state cosmology.

The final blow to the steady-state theory came in 1965 from the work of Arno A. Penzias and Robert W. Wilson of Bell Laboratories. Using a large supersensitive antenna designed to eliminate the usual radio and radar background noise, they detected a weak background radio signal that could not be accounted for by any familiar source. The phenomenon had, however, a cosmological explanation.

In the late 1940s, George Gamow (and in 1965, the Princeton group led by Robert Dicke) had predicted that if the universe had evolved from a fireball state, the very intense high-frequency gamma radiation left over from the explosion would still be around. Due to the continual expansion of the universe, however, the radiation would now be in an enormously red-shifted form. The observed background microwave radiation was the thermal equivalent of about 2.7 degrees above absolute zero (2.7 K), just what theorists had predicted. In 1978, Penzias and Wilson won the Nobel prize for the discovery of this "cosmic background radiation."

One question remains: What happens to the Second Law—the inexorable increase of entropy—if there is enough invisible dark matter and the universe collapses? Will entropy reverse itself—just like a time-reversible mechanical system—and decrease during the stage of contraction, thus getting back to its previous thermodynamic state? Not at all. The Second Law states that all processes increase entropy, whether they are explosions, expansions, or compactions.

Even in our expansionary period of the universe, processes of compactions of matter—similar to a collapsing universe—are occurring, with massive increases in entropy. When two black holes, due to their gravitational attraction, collapse into one hole, entropy increases enormously, as the Bekenstein-Hawking relation shows. Therefore, if the entire universe were to participate collectively in a violent compaction of matter, its entropy would increase tremendously in the process. Thus at the big crunch, the universe will not return to the initial low-entropy thermodynamic state of the big bang, but rather will attain a high-entropy thermodynamic condition.[18] "The final state will be very different from the initial one because of the growth of structure during cosmic evolution," writes cosmologist Craig J. Hogan. "The final state will not be smooth and simple like the initial one but will be bumpy and chaotic, full of turbulence and complexity."[19]

Recent observations of distant exploding stars have shown that the cosmic expansion "is not slowing down as much as many cosmologists had anticipated; in fact, it may be speeding up."[20] If these observations are correct, then the universe appears to be growing at an ever-faster rate—in defiance of all expectations. At the very least, the expansion is not decelerating as fast as once believed. This has created a "revolution in cosmology."[21] Cosmologists are frantically revisiting their theories, which are now in a state of high entropy.

When the Laws of Thermodynamics were discovered, scientists did not know that the universe was dynamic. At the time, they were under the impression that it was a static place. When the evidence came in that the universe is expanding, what changed was our view of how the universe began and how it is evolving. Through observation, scientists are continually refining their theories on the origin and thermodynamic evolution of the universe. The Laws of Thermodynamics, however, remain intact.

Entropy as a Measure of Ignorance and Uncertainty

While Maxwell's demon failed to depose the Second Law of Thermodynamics, it nevertheless made physicists wonder whether we can circumvent the Law of Entropy. In the process came a new revelation that there is a connection between entropy and information. Although Boltzmann, as early as 1894, had remarked casually that entropy is related to "missing information," the physicist who first calculated that relationship was Leo Szilard.[22] In 1929, Szilard published a paper concerning Maxwell's demon problem where he pointed out that the demon couldn't possibly separate the fast particles from the slow ones because it lacks "information" about their velocities.[23] To obtain that missing information, the demon must make an observation. In so doing, the entropy of the total thermodynamic system (demon + gas) increases in conformity with the Second Law.

Boltzmann's relation states that entropy is proportional to the natural logarithm of the number of microscopically different configurations that the thermodynamic system can adopt. As the thermodynamicist D. ter Haar has pointed out, we can express Boltzmann's relation "slightly differently by stating that the entropy measures our *ignorance* or *lack of knowledge* or *lack of (detailed) information*" about the internal arrangements (possibilities) of a system.[24] The greater the number of possibilities, the higher the entropy of the system, and thus the greater is our ignorance or "uncertainty" about its internal configuration.

Let us examine some processes that increase entropy and see how our ignorance increases correspondingly. Assume that we have a molecule moving freely in the left side of a partitioned box. We know where the particle is—in the left side. Suppose we increase the volume by removing the partition, and ask: Where is the particle now? It could be in the left side or the right side. As entropy increased, so did our ignorance about the system—our uncertainty about the particle's location. We will need additional information through observation to determine the particle's position. As the degrees of freedom of individual atoms and molecules increase, so does our ignorance about their state.

Consider instead a thermodynamic system consisting of two nations sharing a common border. The border exists to keep the inhabitants of each nation from intermixing freely, thus preventing entropy increases through the process of mixing. However, when the border fails to act as a barrier and allows free movement between the nations, the entropy of the system increases accordingly. This free mixing deepens the ignorance of the respective governments about the state of their inhabitants.

In recent years, many national borders have practically disappeared as barriers, allowing almost unobstructed movement

of people (or drugs and weapons). Consequently, many governments—including the United States—no longer know the status, or
the address, of a sizable number of people who live and move
around within their territories.

Another way entropy increases in a thermodynamic system is
through an increase in the number of molecules within a given
space. As the molecules increase, more information is required to
describe the system. For example, when world population increased
from 5 billion to 6 billion, we needed more information to keep track
of the additional 1 billion people.

The relationship between increases in entropy and ignorance is
evident in the field of medicine. The medical profession has made
great advances in understanding the human body and our interaction with the environment. But because we live in a thermodynamic world, new viruses and bacteria keep appearing, thus increasing the ignorance of the medical community. In the last three
decades more than 30 new diseases, often virulent, have been discovered. They include Legionnaire's disease, Lyme disease, Ebola,
and AIDS.[25] As new diseases and health disorders arise, the situation calls for new investigations.

The association between increases in entropy and ignorance is
also apparent in the world of chemistry. When we produce new
chemicals, we increase not only entropy but also our ignorance of
our thermodynamic system because we do not know the effects of
those chemicals on plants, fish, animals, and human beings. In the
past three decades or so, we have introduced more chemicals than
in all of preceding history.[26] There are now 65,000 synthetic chemicals used throughout the world. The National Academy of Sciences
(NAS) has pointed out that most Americans are exposed daily to
thousands of chemicals that are "legitimate candidates for toxicity
testing related to a variety of health effects," but "only a few have
been subjected to extensive testing, and most have been scarcely
tested at all."[27] Both NAS and the Office of Technology Assessment
have warned of the unknown dangers of chemical mutagens and
neurological toxicants. "The frightening conclusion is that *we don't
know what we are doing*," writes former State Department official
Lindsey Grant.[28]

The Law of Entropy also has military consequences. For example, Paul Revere did not know the Law of Entropy or informa-

tion theory when he said "One if by land, two if by sea." Yet Revere and his comrades used a concept that emanates from the Law of Increasing Entropy—that is, as the number of ways of attacking the enemy or of being attacked increases, so does the complexity of warfare. Consequently, the commanding officer needs more "bits" of information to communicate to the troops the particular mode of attack. If there were only one way of assailing the enemy, there would be no uncertainty about it, and no need to devise any convention for describing it.

In Paul Revere's time, there were just two modes of attack. Since then, a third has been added—by air. And now, thanks to space age technology, a fourth dimension—space—is being added to the possibilities of attack. Technological advances are enabling humans to bring about more sophisticated and deadlier means of attack and destruction, which calls for more bits of information and constant intelligence gathering.

Irreversible thermodynamic systems, while increasing entropy with the passage of time, generate new information, thus increasing our ignorance about them. Unlike energy, information is not a conserved quantity in thermodynamic systems. It is not possible to backtrack and know everything about a thermodynamic system, nor is it possible to foresee the future.

To see how new information is produced in a thermodynamic system, let us follow the evolution of our universe from the big bang. As odd as it may sound, the model proposes that at the time of the big bang, the universe—all the cosmos that we can observe—was condensed into a region smaller than an atom. Matter as we know it did not yet exist. The universe was pure, exquisitely hot energy in its simplest possible form.

Physics starts at 10^{-43} seconds after the big bang. That infinitesimal fraction of a second is expressed by a decimal point followed by 42 zeros and a 1. It is called the Planck time. In quantum physics, time does not come in smaller portions than this. Gravity had just broken loose from the single unified force—composed of gravitational, electromagnetic, strong, and weak forces—that is presumed to have existed at the big bang. Recently, physicists have predicted that, at ultrahigh energies, the electromagnetic, strong, and weak forces become unified into a single force. In the language of theoretical physics, these three forces are combined under the Grand Unified Theory.

Because physicists do not yet understand gravity well enough, they cannot describe this ultimate unification of forces, and consequently they cannot break beyond the 10^{-43} second barrier. Our universe was then about 10^{-33} centimeters in diameter and was inconceivably hot (about 10^{32} K) but cooling as rapidly as it was expanding. There was no chemistry going on, certainly no biology—only some physics.

At 10^{-35} seconds the strong force that glues together the particles within the nucleus of an atom also began to acquire an independent identity. At this point, only the electromagnetic and the weak nuclear force were still united. Physicists Abdus Salam and Steven Weinberg, working independently, came up with a unified theory of these two forces, which earned them the Nobel prize. This unified force, called "electroweak" force, which can exist only at very high energies, has been verified experimentally by big particle accelerators.

For the tiny universe, an "inflationary epoch" has been theorized, beginning at about 10^{-35} seconds and lasting until about 10^{-33} seconds, in which space expanded at an ever increasing rate—eventually faster than light. Within those split-second inflationary moments, the universe increased its size by about 10^{43} times.[29]

At about 10^{-11} seconds, the universe had cooled down enough and its energy had dropped, allowing the electromagnetic and the weak force to separate. At 10^{-5} seconds after the big event, the universe had grown to about the size of our solar system and had cooled considerably. Matter and antimatter annihilated each other (antimatter is identical to matter but with opposite electric charge). It turned out that there was more matter than antimatter, and the excess is what constitutes all the matter in the universe today.

Between 10^{-6} and 10^{-4} seconds, familiar particles such as neutrons and protons "froze" into existence. When the universe was 3 minutes old, the temperature had cooled enough so that simple compound nuclei could be formed. It took about 300,000 years for the temperature to drop to 3,000 K, allowing electrons to be bound with nuclei, creating full-fledged atoms. This is the point at which matter is said to have "decoupled" from energy.[30] Before this, the universe had consisted of nuclei and electrons swimming in a sea of hot radiation. As the temperature cooled, electromagnetic radiation emerged as photons freed themselves from their bondage with matter, allowing light to travel through space. It took about a billion years before quasars were formed. The observable universe began to assume its familiar appearance: bright points of light in a sea of dark sky.

At the big bang, we could describe practically everything—the entire existing universe—with a few parameters such as diameter, temperature, and perhaps one unified force. Knowing these few facts then, we would know just about everything. However, a few seconds after the explosion, we would already be falling behind in our knowledge because so many things were happening. From Time Zero on, we had to deal with more forces, more particles, more elements, and more interactions. Eventually we had to keep track of such things as stars, solar systems, galaxies, chemistry, ecosystems, living organisms, and biology. It would take enormously more bits of information to describe today's universe and its laws than the universe of the big bang.

Even if we had followed the universe since its inception, and even if we could have observed the universe in its entirety all at once, we would still have been caught off guard innumerable times when new, unanticipated processes and events began to develop. Eventually, we would not be able to keep track of all the information (entropy) generated by the thermodynamic system.

Ever since we invented recording devices to store information, we have been falling behind in our tracking efforts. We cannot even keep pace with our own intellectual inventions (entropy), such as philosophy, logic, languages (especially computer languages), psychology, sociology, and economics. We invented computers with the hope that they would help us catch up with information processing. However, computers have generated such an overwhelming amount of entropy of their own that we cannot even keep up with that entropy.

Humans as an Open Thermodynamic System

Living organisms are engaged in a constant battle to maintain themselves against the forces of entropy and decay.[31]
—Marc Lappé

When physicists first formulated the Laws of Thermodynamics, they considered only isolated systems, which exchange neither energy nor matter with the outside world. This approach works out well for the universe, since the universe includes everything and there can be no "outside world." When energy, but not matter, is exchanged between a system and its surroundings, we have a second type of system—called a closed thermodynamic system. This tends toward a state of equilibrium: the entropy of the system and environment increases toward a maximum value, at which point entropy production ceases. For all practical purposes, Earth and the Sun form a closed thermodynamic system.

Within the thermodynamic family, there is a third and important kind of system—the open system—which can exchange with the environment not only energy but also matter. Life, for example, falls in this category. The entropy formulation can be extended to open systems. Here, we must distinguish two terms in the overall entropy change: The first represents the entropy exchange between system and environment; the second is the entropy produced within the system. The Second Law demands that the entropy production inside the system be positive.[32]

In purely scientific terms, humans are a thermodynamic system obeying the Laws of Thermodynamics. We take in energy (Q) from our environment in the form of chemical energy (food and oxygen) and perform work (W) by walking, running, or doing other physical activities. In accordance with the First Law, if the overall energy balance ($Q - W$) is positive, we put on some weight. If we become overweight, we can lose weight by taking in fewer calories, or doing more work, or both.

Although human beings are often referred to as a heat engine, we are in many ways more analogous to a fuel cell, which is a constant-temperature thermodynamic device. Besides doing work, we strive to maintain a constant temperature. Therefore, the energy released through chemical reactions—combustion of food with oxygen—is used not only in doing work, but also in transferring heat to the environment.

The First Law is certainly applicable to the human body. During a given day, if the calories produced by the oxidation of food in the body exceed what flows out, the body will store the excess calories, mainly as fat. An average adult in good health has about 75,000 Calories of stored energy in his or her body. Even if we perform no work at all, lying quietly in a comfortable warm room, we still give off heat at the rate of 43 Calories per hour—about 1,000 Calories per day.[33]

Living organisms are indeed open thermodynamic systems, exchanging both energy and matter with their environment. The growth, maintenance, and decay of all living systems (plants and animals) constitute an open system. Open systems are not in thermodynamic equilibrium by the very fact that energy and matter are being exchanged, with entropy produced in the process.

However, a macrosystem that is not in thermodynamic equilibrium will, in accordance with the Second Law, spontaneously increase its entropy and slowly move toward more disordered states and decay toward equilibrium. Therefore, a constant expenditure of work (entropy production) is required to avoid the rapid decay into an inert state of equilibrium, and to restore the macrosystem to its proper state. The living organism avoids the decay by eating, drinking, and breathing, and in the case of plants, by assimilating. This process is called metabolism, from the Greek word meaning "change or exchange."

This exchange involves taking in highly ordered structures such as food and, through a series of metabolic processes, breaking them down into relatively simple molecules and then building them up again into highly complex organisms such as human cells. A living cell can be viewed as a biochemical factory. Thousands of reactions occur continually in a carefully orchestrated fashion. Large molecules are broken down and others are synthesized through a complex sequence of operations.

The human body undergoes many cyclical processes. Cells decay and are replaced by new ones. Tissues and organs continually replace their cells in repeated cycles. For example, the pancreas replaces most of its cells every 24 hours, the stomach lining is replaced every 3 days, and white blood cells are turned over every 10 days. This mechanism of self-renewal is often called self-organization.[34]

Unlike frictionless mechanical systems, open thermodynamic systems cannot operate without some increases in entropy. Consequently, the cyclical processes occurring in living organisms—like other ecological cycles in Nature—advance with some entropy production. At the end of each cycle, neither the system nor its surroundings return exactly to their initial state; every decayed cell does not get replaced by an identical one. Certain cells in the body rarely, if ever, reproduce and renew themselves—like the ones in the kidney, heart muscles, and the spinal cord; these cells grow old with the individual.[35]

When a living organism is perturbed, its entropy production increases. But the system reacts by returning to the state at which its entropy production is minimum. If our body temperature rises, the rate of deterioration and decay increases as the denaturation of macromolecules increases. Within a certain operating range, the body reacts to bring the temperature down, and thus minimize entropy production. This process is called negative feedback. If the temperature exceeds that operating range, the system goes out of control—positive feedback. At high enough temperatures, the body cannot dissipate the heat fast enough and the metabolic rate increases, producing even more heat and causing the chemical reactions to go even faster, thus giving rise to more heat. This runaway feedback eventually leads to death as entropy production explodes.

Similarly, when the body temperature becomes too low, the biochemical reactions do not proceed at a rate fast enough to provide the necessary energy for the replenishment of the decayed macromolecules and the maintenance of life. Even at low temperatures, molecular disorganizing changes occur; that is, entropy increases within the body. If the body fails to replace the denatured proteins, the body decays rapidly, resulting in the death of the organism.

A living organism, then, constantly fights the law of molecular disorder by continually doing work to maintain its structural organization. In his book *What is Life?* physics Nobel laureate Erwin

Schrödinger wrote: "Every process, event, happening—call it what you will; in a word, everything that is going on in Nature means an increase of the entropy of the part of the world where it is going on. Thus a living organism continually increases its entropy—or, as you may say, produces positive entropy—and thus tends to approach the dangerous state of maximum entropy, which is death."[36]

Schrödinger asked the basic question, How does a living organism stay aloof from death? It does so by continually drawing "orderliness" from its environment to compensate for the entropy increase it produces by living. Schrödinger coined a new term—"negative entropy"—for order. He did it by playing a mathematical trick, which physicists love to do. In Boltzmann's equation, if Ω "is a measure of disorder," its reciprocal, $1/\Omega$, "can be regarded as a direct measure of order," Schrödinger said. Since the logarithm of $1/\Omega$ is minus the logarithm of Ω, he rewrote Boltzmann's equation in this form: $-(entropy) = k \ln (1/\Omega)$.

"Hence the awkward expression 'negative entropy' can be replaced by a better one: entropy, taken with the negative sign, is itself a measure of order."[37] The expression negative entropy, or "negentropy" as it became known, is not only awkward but also misleading. As mentioned earlier, physicists measure the disorder of a thermodynamic system by the number of microscopically different arrangements (Ω) that the system can adopt. At the very least, a system can configure itself one way ($\Omega = 1$) which, in physics, is considered to be in perfect order ($entropy = k \ln 1 = 0$). No system can arrange itself a fractional way; $1/\Omega$ has no physical meaning. As the chemist Henry A. Bent reminds us, "$k \ln \Omega$ *is never negative. A system (to be a system) must possess at least *one* micromolecular configuration."[38]

Mathematically speaking, what Schrödinger did is to change the sign on both sides of Boltzmann's equation. While Boltzmann's entropy relation tells us that entropy is always a positive number, Schrödinger's entropy equation tells us that the negative of entropy is always a negative number (because $k \ln (1/\Omega)$ is always negative).

What is more important is that humans, or living organisms in general, cannot "reverse" entropy by taking in more food than they have produced while living. The metabolic processes that keep us alive in themselves generate entropy. The process of living involves the continual increase in entropy driven by the Second Law of Thermodynamics. As Nobel laureate in chemistry Ilya Prigogine points out, "life is associated with entropy production and therefore with irreversible processes."[39]

To compensate for this dangerous downhill movement, the living organism is constantly engaged in a fight to recoup loss in orderliness. The battle for survival entails taking in from the environment highly ordered substances—low entropy rather than the misleading word negentropy—and using them to maintain orderliness. The organism then releases back to the environment substances in a much-degraded form—high entropy. It is this exchange of low entropy for high entropy that keeps living organisms alive and evolving.

Why Do We Age Irreversibly?

Everything in the universe ages, including the universe.[40]
—Leonard Hayflick, *Cell Biologist*

A process which in no manner can be completely reversed I called a "natural" one. The term for it in universal use today, is: "Irreversible."[41]
—Max Planck

To date, there is a proliferation of conjectures on why living organisms age and eventually die. A synthesized, scientific theory of aging does not exist. In 1983, when gerontologist Dr. Roy Walford was asked whether we know what causes people to age, he replied: "Nobody knows the exact cause."[42] And in a 1997 article entitled "Why Do We Age?" *U.S. News & World Report* reiterated the assertion that "No one yet knows what causes aging and its inevitable consequence, death."[43]

There are many definitions or criteria of what constitutes bio-logical aging. But they are all based on the general postulate, spe-cifically stated or implied, that "there exist gradual changes in the structure of organisms which are not due to preventable diseases or other gross accidents and which eventually lead to the increased probability of death of the individual *as he grows older* with the pas-sage of time," notes biologist Bernard Strehler in *Time, Cells, and Aging*.[44]

Biological aging, then, has to do with gradual changes in the structure of organisms. Although we do not necessarily notice the process at every moment, our bodies undergo gradual entropic changes minute by minute, hour by hour, and day by day. As time goes on, we begin to see and feel the cumulative effects of entropy increases within us. One of the biggest compliments we can receive from a person who has not seen us for a while is, "You look the same; you haven't changed a bit." We are elated that our internal entropy changes are not noticeable from the outside. But deep in our hearts we know we have aged. We are now on some medica-tion, we have replaced a couple of teeth with a permanent bridge, we had our appendix removed, we wear contact lenses as our eyes have deteriorated, we can no longer run and swim as fast as we used to, and our reflexes have slowed down despite a regimented exercise program.

In thermodynamics, the process of aging has no hidden mys-teries. Living organisms age with the passage of time because their entropy increases. In fact, increases in entropy and the passage of time are two sides of the same coin. There would be no passage of time if there were no increases in entropy, as each system would—like a frictionless pendulum—cyclically return to its previous state only to start a new, identical cycle. We say that the universe is aging with the passage of time because all of its constituents are increas-ing their entropies. Aging and increase in entropy are practically synonymous.

In the literature, the Law of Increasing Entropy is rarely ac-knowledged as the essence of aging. Strehler, however, mentions it a few times in his book. He writes: "Entropy will increase sponta-neously in any macrosystem which is not already in its most stable condition and which possesses sufficient ambient thermal energy to permit reactions to occur. Aging and most other disorganizing changes, of course, would not take place at absolute zero because change involves the motion of molecules, their parts, and their aggregates."[45]

Living organisms fight the Law of Entropy by continually expending energy to repair its disorganizing effects. In so doing, they gradually increase their entropy. Eventually, they succumb to their internal disorders and die. To survive, species engage in an activity called reproduction, in which they replace the whole organism. For species, even reproduction has not proved fail-safe: Many species have bowed to the Second Law and have become extinct, having reached their state of maximum entropy.

If we cannot stop the aging process, can we reverse it? Is that a scientific possibility? According to the Second Law, it is not. In the physical sciences, "irreversibility" has a distinct meaning, as explained by Max Planck: "That a process may be irreversible, it is not sufficient that it cannot be directly reversed. This is the case with many mechanical processes which are not irreversible. The full requirement is, that it be impossible, even with the assistance of all agents in nature, to restore everywhere the exact initial state when the process has once taken place."[46]

The central point of Planck's definition lies in the requirement that we need to restore a system, in its entirety, to an earlier state. Reversible (mechanical) systems can, at least conceptually, return to their earlier state, but it is impossible to return an irreversible system back to its initial state. The Second Law is based on the fundamental axiom that all natural processes are irreversible. Consequently, in thermodynamics, "natural processes" and "irreversible processes" are the same.

The question whether biological aging is a reversible process boils down to asking, Can we transform ourselves, "body and mind," in totality, to an earlier state? As Prigogine points out, this is impossible. "In classical dynamics, some simple situations that are time reversible ($t \leftrightarrow -t$) can at least be conceived of," states Prigogine. "Whenever chemical processes (and a fortiori whenever biological processes) are considered this becomes impossible because chemical reactions are always associated—nearly by definition—with irreversible processes. Moreover, measurements—which extend our sensory perceptions—necessarily involve some element of irreversibility."[47]

In spite of many predictions that agelessness and immortality for humankind are just around the corner, all of us are still aging irreversibly.[48] We cannot stop aging because, quite simply, we have within us the inexorable increase of entropy dictated by the Second Law.

Is Evolution a Miracle in Violation of the Second Law?

Evolution is often described as a process that runs counter to the Second Law of Thermodynamics—and hence a miracle. In *The Road Less Traveled*, for example, M. Scott Peck writes: "The most striking feature of the process of physical evolution is that it is a miracle. Given what we understand of the universe, evolution should not occur; the phenomenon should not exist at all. One of the basic natural laws is the second law of thermodynamics, which states that energy naturally flows from a state of greater organization to a state of lesser organization, from a state of higher differentiation to a state of lower differentiation. In other words, the universe is in a process of winding down."[49]

The explanation given by Peck—as to why evolution is a miracle—is a common one:

> The process of evolution has been a development of organisms from lower to higher and higher states of complexity, differentiation and organization. A virus is an extremely simple organism, little more than a molecule. A bacterium is more complex, more differentiated, possessing a cell wall and different types of molecules and a metabolism. . . . Insects and fish have nervous systems with complex methods of locomotion, and even social organizations. And so it goes, up the scale of evolution, a scale of increasing complexity and organization and differentiation, with man, who possesses an enormous cerebral cortex and extraordinarily complex behavior patterns, being, as far as we can tell, at the top. I state that the process of evolution is a miracle, because insofar as it is a process of increasing organization and differentiation, it runs counter to natural law. In the ordinary course of things, we who write and read this book should not exist.[50]

In the literature, and consequently in the popular mind, misconceptions about the Second Law abound. The law states unequivocally that entropy increases in all natural processes irreversibly—irrespective of whether the bodies involved in the processes are higher, lower, or identical in complexity, differentiation, and organization. Moreover, the law does not state that all natural processes tend to produce less organized, less differentiated, simpler molecules, or the opposite; if it did state either one, it would be a sure loser, since both phenomena occur in Nature and, of course, neither one violates the Second Law.

Consider the nuclear processes occurring naturally in the Sun and other stars. One of the chain reactions involves the fusion of four protons—H^1 (mass 1) nuclei—to form a more complex nucleus helium 4. The production of one helium nucleus from four protons is an example of a process called fusion, in which a heavier (more complex) element is built up from lighter (simpler) elements. In the process, subatomic particles are emitted, along with heat and electromagnetic radiation.

Similarly, the radioactive decay of heavy particles occurs naturally. In radioactivity, a heavy (complex) element disintegrates into lighter (simpler) and more stable elements, emitting very light radioactive elements and radiation. The point is that in Nature, we find both processes—fission and fusion—occurring naturally. In fission, complex elements disintegrate into simpler elements, while in fusion, simple elements combine to form complex elements. Neither fission nor fusion violates the Second Law. Both processes have been in action naturally for billions of years, increasing the entropy of the universe.

Astrophysicists tell us that our universe began with the big bang. At the beginning there were no complex elements like oxygen, iron, uranium, and human beings; there was no chemistry or biology going on, only some physics. Just after the big bang, there were only elementary particles and much heat. Since then, more complex elements have appeared, as well as the highly dissipated cosmic background noise.

Thermodynamics warns us not to draw general conclusions by examining part of a thermodynamic system. Consequently, the observation that more complex, more structured, and more organized elements have evolved from simpler, less structured, and less organized entities does not tell us anything about the overall entropy trend of the universe, nor does it suggest that the phenomenon it-self runs counter to the Law of Increasing Entropy. In fact, if we cal-culate the entropy of some monatomic gases using Boltzmann's equation ($S = k \ln \Omega$), we find that the entropies of the elements in-crease with increasing atomic weight—that is, with increasing levels of complexity, structure, and organization. The same situation pre-vails for diatomic and triatomic molecules.[51]

Is the evolution of the universe, from the big bang to the present, in violation of the Second Law? Not at all. The entropy of the uni-verse has been increasing through such processes as expansion (in-crease in volume and dispersion of energy-matter) and energy transformations of various kinds. Does biological evolution violate the Second Law? Of course not. As indicated earlier, our system is composed of Earth and the Sun. Earth has been receiving energy from the Sun for billions of years in the form of radiation and heat. It is this energy flow that has allowed life to evolve.

Physical, chemical, or biological evolution cannot be in con-tradiction to the Law of Entropy. In fact, in 1865 Rudolf Clausius specifically called the quantity "S" entropy because it is about trans-formations—in Greek it means "turning into." The "miracle" of evo-lution did not start with the appearance of life on Earth, a few billion years after the start of the universe. Rather, the miracle began with the creation of the universe, and is still going on.

We exist today because of the Second Law. If the universe were created as a lifeless reversible mechanical system, where entropy never increased, novelties would have never appeared; we would never be around. We are alive and evolving because of the Law of Entropy.

Chapter

Knowledge and Ethics

Incoherence is the result of the present disintegrative processes in education.[1]

—Ruth Nanda Anshen

Knowledge Undergoes Thermodynamic Transformation

Where is the wisdom we have lost in knowledge?
Where is the knowledge we have lost in information?[2]
—T. S. Eliot

In everyday life, modern humans expend a substantial amount of energy in the acquisition and dissemination of knowledge. Concerning Maxwell's demon, Leo Szilard and Dennis Gabor formulated the idea that knowledge and information cannot be obtained without the expenditure of energy, thus without some increase in entropy. Observation, acquisition, retention of scientific data, and the dissemination of results all demand some work. In the process, work is transformed into knowledge. Unlike the Sun's energy, which we receive every day from the heavens, we do not get a daily dose of free knowledge from outer space. Consequently, we must expend energy to acquire, maintain, and organize our stock of knowledge.

Like energy, knowledge also takes many forms. Some knowledge is fundamental and low-entropic in the sense that it hardly dissipates through time. The basic laws of physics fall into this category. Once discovered, tested, and verified, they are rarely—if ever—overthrown. There are times a law is found invalid when applied to new areas of investigation. In that case, a more general law or theory is formulated to extend the area of applicability.[3]

Knowledge of the Laws of Nature is important and fundamental. However, even if we knew all of Nature's Laws, we would still be ignorant of the state of a given system. For example, Kepler's laws of planetary motion do not tell us anything about the positions and velocities of the planets. For that we need observations, measurements, and calculations. These activities require energy expenditure, leading to an increase in entropy.

Currently, humankind is devoting a considerable amount of energy to gathering information about a wide spectrum of thermodynamic systems, from tiny viruses, bacteria, and insects to planets, galaxies, and the universe at large. But we are not merely learning about the Laws of Nature and passively observing the events in our universe. We are actively applying our acquired knowledge to develop and produce such things as technologies, chemicals, drugs, and living organisms. Consequently, our information system, or knowledge base, includes our own activities.

As we pump energy into the information processes, knowledge expands in volume and becomes more complex, and in turn absorbs more energy from us. As time goes on, more effort is required to push knowledge to its next level of complexity. We combat this situation by increasing the number of scientists and people of intellect. In 1797, when Thomas Jefferson became the president of the American Philosophical Society, all American scientists and their colleagues in the humanities could be seated in the lecture room of Philosophical Hall. Today, their successors include "450,000 holders of the doctorate in science and engineering alone."[4] It takes more scientists now to keep the flames of knowledge burning.

During his life, Sir Isaac Newton invented differential and integral calculus, discovered the universal law of gravitation, and made numerous contributions in chemistry, astronomy, mathematics, and physics, yet still had time left for philosophy. By the time Albert Einstein was born—about a century and half after Newton's death—the problems in physics had taken on a different aspect. They had become a lot more complex and far more mathematical. The balloon of knowledge had swollen to such a magnitude that no person, however brilliant, could make major contributions in many fields of knowledge. The opposite was happening: A number of high-powered mathematicians were assisting Einstein in smoothing out the rough edges of his relativity theory. And during the last 25 years of his life, Einstein devoted considerable effort—without much success—to trying to develop a unified field theory.

As time goes on, we inject more and more energy into a system, but we get less and less out because the thermodynamic system becomes increasingly inflated. Economists call this phenomenon the

law of diminishing returns. Whenever the entropy of a thermodynamic system increases, so does our ignorance about it. In the case of
knowledge, we struggle against complexity and high entropy by
specializing—by devoting attention to narrower and narrower areas
of investigation. This in turn leads to a situation where communication between people of knowledge becomes increasingly difficult.

As physicists invaded deeper and deeper into Nature's inner
workings, the mathematics they needed became increasingly complicated and demanding. Consequently, physicists now have great
difficulties explaining their work to others, as well as following the
work of other scientists.[5] "Every one of our laws is a purely mathematical statement in rather complex and abstruse mathematics.
Newton's statement of the law of gravitation is relatively simple
mathematics. It gets more and more abstruse and more and more
difficult as we go on," Richard Feynman once observed.[6] Today
physicists have to juggle an increasing number of complicated laws
and theories, and cut through layers and layers of complexity—
theoretical and experimental—before they can see the physics behind what they are doing.

Knowledge is behaving like the universe: evolving from a state
of concentration and simplicity to a state of diffusion and complexity, from a state of low entropy to a state of high entropy. For example, the laws of physics were once concentrated into Newton's
laws of mechanics. But then came electromagnetic phenomena,
heat and thermodynamics, statistical mechanics, atomic and nuclear physics, quantum mechanics and relativity, elementary particle physics, and more recently string theory—a theory that could
possibly unify all four forces of Nature, Einstein's dream, into what
is called "Theory of Everything."

Physicists who are knowledgeable in string theory point out that
it is intricate and difficult to follow even for physicists. In the words
of Steven Weinberg: "String theory is very demanding; few of the
theorists who work on other problems have the background to understand technical articles on string theory, and few of the string
theorists have time to keep up with anything else in physics, least of
all with high-energy experiments."[7] In *The Elegant Universe*, string
theorist Brian Greene writes that "the mathematics of string theory
is so complicated that, to date, no one even knows the exact equations of the theory. Instead, physicists know only approximations to

these equations, and even the approximate equations are so complicated that they as yet have been only partially solved."[8] Physics has
indeed reached a high level of complexity.

Laws, theories, and ideas do evolve into more diffused states.
For example, when only the laws of mechanics were known, physicists' attention was focused on them. As new principles of physics
were discovered, classical mechanics and Newton's laws of motion lost their overall importance: they were degraded, becoming
part of a bigger and more complicated system of physical laws and
theories.

The degradation and diffusion of ideas occur in all fields of
knowledge. In philosophy, for example, people's attention was once
focused on just a few philosophical points of views. As time went
on, more and more philosophers and people of intellect emerged,
each suggesting a few ideas, or criticizing other viewpoints, or putting forth a few paradoxes. Thus with the passage of time, humanity piled up a dizzying network of philosophical ideas, opinions,
and paradoxes.

Today, a particular idea no longer makes a great impact on us
because it becomes diffused among so many others. As entropy in
knowledge increases, the system becomes noisier. Consequently, to
get people's attention, we need to shout louder and more often.
With the passage of time, the accumulated pile of concepts, events,
theories, laws, canons, and paradigms puts a greater and greater
load on the human mind. Deciding whose literary work, philosophy, or ideology to study, follow, or read becomes a monumental
undertaking. Humanity has to keep track of and sort out an ever-
increasing heap of ideas. Following historical events also becomes
increasingly burdensome. Studying just the historical happenings
since World War II would consume considerable time.

In all fields of knowledge, professors and researchers are publishing their findings at a maddening pace. Mathematicians publish
an estimated 200,000 theorems every year. Not surprisingly, most of
them are ignored.[9] This rapid expansion of material—entropy—produced by humanity puts a tremendous drag on those who have to
keep up with what is going on. "By now it is impossible for a theoretical physicist to read all the papers even in some narrow subspe-
cialty," writes Weinberg, "so most articles on theoretical physics
have little impact and are soon forgotten."[10]

The concept of irreversibility of processes also applies to the knowledge we accumulate. As knowledge expands, its entropy increases irreversibly. There is no way that we can put our knowledge base—in its totality—back to what it was yesterday, let alone what it was a century ago. We cannot unlearn the knowledge that $E = mc^2$, nor can we hide the technology that converts mass to energy. Consequently, every item of knowledge we acquire, especially about the inner workings of Nature, adds a certain responsibility as to its usage, because its applications will irreversibly change the state of our thermodynamic system—from a lower to a higher entropic state. As time goes on, the applications of knowledge demand from us greater and greater wisdom.

Thermodynamic View of the Educational System

Societies, like everything else in the universe, undergo transformations. After a time, they tend to become more complex. In modern times, the educational system has played a major role in transforming societies from simpler entities to more complex ones. The educational system takes in simple, low-entropic human beings and feeds them increasingly complex ideas, procedures, and theories, transforming them into high-entropic, complex human beings. When the graduates leave the educational system, they come into society and transform it into a more complex entity. Simple procedures are replaced with more complicated ones. Gradually, society becomes more complex as it produces and absorbs more college graduates with higher and higher academic degrees. Entropy increases as complexity mounts.

Consequently, society needs even a greater number of educated people who can function in the new, complex environment. The educational system produces lawyers who write laws, which in turn require more lawyers to interpret them and deal with them; the system generates economists who come up with socioeconomic theories and processes, necessitating more sociologists and economists to consider their effects on the economy and on society.

The more energy we pump into the educational system, the more it expands and the more complex it becomes. As the output of the educational system enlarges, society's entropy increases accordingly. Not surprisingly, the United States, which leads the world in the number of college graduates, has attained a level of complexity that no other society can even approach—from its government regulations, computers, and communication and military systems to its legal, financial, health, and tax systems.

In addition, educated people create tools, machinery, and technology that increase not only the complexity of the world we live in but also the capability to produce even more complex tools, machinery, and technology. For example, for a long time we could identify engineers on college campuses by the slide rule they wore on their belts. A slide rule is a simple but amazing calculating device costing no more than a few dollars. When engineers got out of school, they were fully functional with a slide rule, a tool that would last a lifetime. With the help of this simple tool they designed bridges, houses, automobiles, radios, televisions, trains, and airplanes.

But a group of physicists, engineers, and mathematicians invented a more complex tool—the computer. Little by little, the durable slide rule became worthless; it was replaced by the computer. Now engineers have to be proficient not only in their own engineering field but also in complex computer technology and languages. Thus, a greater effort is demanded from engineers. And when they graduate and leave the university compound, they are not functional without a computer. Society has to supplement their acquired knowledge with massive computing power—a substantial increase in entropy.

First, scientists and engineers became dependent on computers. Then other professions gradually began to join the group. Economists, sociologists, political scientists, and analysts started to use computers to generate more and more numbers, or "relevant statistics." Through data obtained from surveys, financial records, personnel rolls, opinion polls, and the like, they began to build complex social models: "econometrics" in economics, "cliometrics" in history, and "psychometrics" in behavioral sciences.

As distinguished sociologist S. M. Miller has pointed out, "In metrics, you can use lousy data on all kinds of assumptions and get very precise and misleading results," adding that "computers are one of our great disasters—if you multiply data at an enormous clip, you have very few ways of analyzing it. A famous study . . . came up with the most meaningless results you could imagine. Why? They threw every variable into the computer and cross-tabulated it. Fifty thousand correlations overwhelmed them—they had no sense of what was going on."[11]

In a similar situation, President Reagan's budget director, David Stockman, openly admitted that during the preparation of the nation's budget the bewildering set of numbers coming out of the computers confused even those who produced them. "None of us really understands what's going on with all these numbers," Stockman confessed. "The whole thing is premised on faith. On a belief about how the world works."[12]

First, the grown-ups became nonfunctional without computers. Today, almost everybody—including young people, the future generation—has become dependent on computers. American kids are surrounded with more calculators and computers than ever before. Computers are being used not only to perform mathematical calculations but also to teach how to calculate.

Despite the prevalence of computers in American homes, classrooms, and colleges, however, scientific knowledge remains in the background. In 1984, Nobel laureate Salvador E. Luria pointed out that we live in a society where there is much talk about science, yet only a small percentage of the people is "equipped by schooling, including college, to understand scientific reasoning."[13] In 2000, a report by the National Science Foundation confirmed Luria's assertion: "Most Americans know a little, but not a lot, about science and technology."[14] Gary Chapman, director of the 21st Century Project at the University of Texas at Austin, adds in the *Los Angeles Times*: "Given some of the findings, even that may be generous."[15]

Luria talked about the consequences of U.S. "Ignorance of Science." He remarked: "There are a lot of kids who know everything about computers—how to build them, how to take them apart, how to write programs for games. But if you ask them to explain the principles of physics that have gone into creating the computer, they don't have the faintest idea."[16]

It does not hurt to remind ourselves that nearly all the major laws and principles of physics—from Newton's law of gravity and the Laws of Thermodynamics to Einstein's $E = mc^2$ and Heisenberg's uncertainty principle—were discovered when the computer was not around. In making these discoveries, scientists used pencils, paper, and their brains.

Luria's remark applies equally well to adults. Many educators, including some computer scientists, have warned young students not to be mesmerized by computers and computer programming.[17] Scientific knowledge does not reside in human-made computer languages or in the complicated technical manuals that teach us how to use these machines.

In today's highly complex, technological world, ignorance of science has wide-ranging implications. Luria points out that

> the failure to understand science . . . leads to a blurring of the distinction between science and technology. Lots of people don't distinguish between the two. Science is the production of new knowledge that can be applied or not, while technology is the application of knowledge to the production of some product, machinery or the like. The two are really very different. . . . Science, in itself, is innocuous, more or less. But as soon as it can provide technology, it's not necessarily innocuous. No society has yet learned how to forecast the consequences of new technology, which can be enormous.[18]

Basic science provides us with knowledge and principles of how Nature works. It generates relatively little entropy. Technology—the massive application of knowledge—produces a tremendous amount of entropy and breeds complexity in our thermodynamic system.

Disorder in Knowledge

It isn't so much the ignorance of mankind that makes them ridiculous as knowing so many things that ain't so.[19]
—Josh Billings

Our knowledge—what we have recorded in our brain, books, magazines, newspapers, publications, and computers—includes everything we know about Nature, about ourselves, and about our inventions. Like other natural processes, our knowledge obeys the Law of Increasing Disorder.

With the passage of time, we accumulate more and more bits of incorrect information. Here is an example: A newspaper prints some information, which is later discovered by the newspaper or a reader to be incorrect. The next day or so, the editor prints a correction, the only recourse the editor has for a mistake. But the damage is already done. There is no assurance that the people who read the first article (or were informed about it through secondary sources) will necessarily read its correction and clear the misinformation from their brain. More important, incorrect information does not always get corrected. This happens, of course, with newspapers, magazines, books, computers, radio, television, and all other media that record and disseminate information. Thus, disorders in knowledge multiply and propagate.

Our scientific textbooks are no exception. They have become so littered with factual mistakes, they misrepresent the essence of science, *Newsweek* has noted.[20] The blunders range from mere oversimplifications to howlers so bad that Newton would have difficulty recognizing his laws. Many popular misconceptions about scientific laws come from "facts" learned from books. Moreover, the disorders are propagating "like maggots," because most new textbooks are closely modeled on profitable predecessors.[21]

Inconsistencies and misinformation take on many forms and are reproduced in many ways. For instance, it is not uncommon to watch on television or read in newspapers Nobel laureates in economics propounding contradictory economic theories. They have received the highest honor that humankind bestows on their profession. Yet the millions of people who listen to them, or read their theories, receive and store in their brain inconsistent messages. It could be that the assumptions behind their economic variables are different, or that they are using different economic models. Whatever the case, the effect is the same: a substantial entropy increase in knowledge.

In our information base, we have another troublesome kind of disorder. Some of the basic scientific facts published in scientific journals are incorrect—not through human error, which is quite possible, but through falsification by scientists and researchers. "Although the foundation of science is trust and its goal is truth," wrote *Newsweek*, "the field has been rocked by scandal dating back at least to Piltdown man, the faked fossil skull discovered in 1912. More recent frauds have triggered a wave of soul-searching over what constitutes fraud and why it occurs."[22] The scandals involved the faking of research results in some of the most prestigious universities, laboratories, and hospitals, particularly in the field of biomedicine.

In the United States, incidents such as the following have contributed to the disorder in knowledge. At a medical school, a fraud case involved the testing of heart drugs on dogs. When the finding that two drugs limited damage after a heart attack was challenged because it contradicted results from other labs, the researcher could not substantiate his work. It turned out that the scientist had "fabricated data in 109 publications." In the most celebrated case, a scientist at a cancer research center darkened skin on a white mouse with a pen to fabricate evidence that skin grafts from another species of mice could be successfully done. The achievement was hailed as a major breakthrough in cancer research, until the fraud was unveiled. As one scientist observed, "much of what is published goes unchallenged, could be wrong, and the chances are that nobody knows."[23]

In other cases, incorrect scientific conclusions were drawn from data that had been faked. When fossil hunters in England unearthed Piltdown man—which was just fragments of a large brain case and an apish jaw—they claimed it showed that a big brain was the first human feature to emerge in the evolutionary processes.

However, anatomical and chemical tests in 1953 revealed that the Piltdown pieces came from different creatures. The jaw was a modern orangutan's.[24]

Scientific fraud has been the subject of numerous articles in scientific journals and even investigations by the U.S. Congress. Two books, *Betrayers of the Truth* and *False Prophets*, provide ample evidence that this disorder is widespread and that it has been around since the beginnings of science.[25] Fraud is even more of a problem in the social and biological sciences. Because of uncontrollable variables in these sciences, the experiments are hard to reproduce; thus fraudulent results have a better chance of being concealed.[26]

In our knowledge base, we have yet another disturbing kind of disorder. When we search the literature about a certain scientific principle, we find vast inconsistencies in interpretation. In subjective fields, we expect to find differences of opinion on a given issue. However, when we read about scientific truths or Laws of Nature, we do not expect to find wide-ranging opinions about them. Yet we do.

The literature, for example, is full of references to the First and Second Laws of Thermodynamics. What do people say about these laws? Here are some samples. In his 1996 book, *The Ultimate Resource 2*, economist Julian L. Simon devoted a section on entropy and the Second Law, where he stated unequivocally, "The concept of entropy simply doesn't matter for human well-being."[27] Many people have read this book. Some may even have told their friends about their newly acquired information. All those individuals are under the impression that the Second Law of Thermodynamics has no effect on their lives.

A reference to the Second Law can also be found in George Gilder's *Wealth and Poverty*, published in 1981. He writes: "The real world, say the entropy theorists, is a world of irreversible time, governed by the second law of thermodynamics: the entropy law—the tendency of energy (*negentropy*) to dissipate irretrievably into entropy as it is used."[28] Gilder then gives his opinion on the subject: "But we should chiefly listen not to the words but to the music and acknowledge that despite the pretense of science, entropy theory is essentially a metaphor."[29] Physicists have tried to reverse the irreversible trend of natural processes, and thus beat the Second Law, but could not do it even with the help of imaginary demons. But

Gilder has reduced the unshakeable Law of Entropy to a mere metaphor by a stroke of the pen.

Gilder's book received much attention in the communications
media, as well as in economics and political circles. The book even
made it to the President of the United States. In March 1981, Associated Press captured a photo that was widely picked up by the afternoon dailies. It showed President Reagan visiting the hospital bed of
the ailing Senator Bob Dole and presenting him with Gilder's *Wealth
and Poverty* as a get-well gift.[30] Entropy is discussed in the last chapter. All those who have read it are under the impression that the Law
of Entropy is just a metaphor and has no effect in our daily lives.
Then we wonder sometimes why Representatives, Senators, and
even Presidents make statements or pursue policies that are contrary
to the Laws of Nature and everyday experience. The reason lies in
the fact that our information system, which we call knowledge, is in
a highly disordered state—that is, in a high-entropic state.

Consider another example of informational disorder. It comes
from R. Buckminster Fuller, the noted architect and technologist. In
his 1981 book *Critical Path,* he wrote:

> We have two fundamental realities in our Universe—the physical
> and the metaphysical. Physicists identify all physical phenomena
> as the exclusive manifest of energy: energy associative as *matter* or
> disassociative as electromagnetic behavior, *radiation*. Both of these
> energy states are reconvertible one into the other. Because there is
> no experimental evidence of energy being either created or lost,
> world scientist-philosophers now concede it to be in evidence that
> Universe is eternally regenerative.[31]

The creator of the geodesic dome must have been telling his
readers about the First Law of Thermodynamics, about the conservation of energy-matter. Physicists have been saying since the
middle of the nineteenth century that all processes are irreversible,
the crux of the Second Law. Physicists do not claim that cosmic
background radiation is convertible into gravitational matter, like
iron, gold, and titanium. They also do not claim that the universe
is eternally regenerative.

Interestingly, Fuller compiled in the same book a "Chronology
of Scientific Discoveries and Artifacts." There, for the year 1847, the
table has two entries, one of which is the "Second law of thermo-

dynamics."[32] The question remains, How can radiation and matter be converted back and forth endlessly within the Second Law? It cannot.

Ironically, Fuller wrote his book on "how mankind *can* survive," because of his "driving conviction that all of humanity is in peril of extinction if each one of us does not dare, now and henceforth, always to tell only the truth, and all the truth, and to do so promptly—right now."[33] The law of conservation of energy-matter is part of the scientific truth but not all the truth. The law stating that entropy increases in all energy transformations is certainly more fundamental to the survival of humankind than the law stating that energy cannot be created or lost.

It is impossible to ignore the highly disordered state of our knowledge base. Fuller was upset about all the misinformation around us. He thought, however, that computers would solve the problem in due time: "Computers will correct misinformed and disadvantaged conditioned reflexes, not only of the few officials who have heretofore blocked comprehensive techno-economic and political evolutionary advancement, but also of the vast majorities of heretofore-ignorant total humanity."[34]

If we took Fuller's advice and put in the computer his statement that the universe is eternally regenerative because there is no experimental evidence of energy being either created or lost, or if we put in Gilder's statement that entropy theory is essentially a metaphor, or Scott Peck's assertion that evolution runs counter to the Second Law, and then broadcast the information through the Internet to educate the heretofore-ignorant masses, all we will be doing is spreading informational disorder at nearly the speed of light.

Where does the disorder in knowledge emanate from? The answer was given in 1957 by William Oliver Martin: "Disorder of knowledge takes its beginnings in higher education, and there it must be corrected."[35]

"The basic characteristic of Man," wrote Pierre Teilhard de Chardin, "the root of all his perfections, is his gift of awareness *in the second degree*. Man not only knows; he knows that he knows."[36] While it may be meaningful to be aware that "we know that we know," it is much more important to know the state our knowledge base is in. What we ought to be cognizant of is that our knowledge is not only in a state of high entropy—disorder—but that its entropy is increasing rapidly.

Chapter   5

The United States in High Entropy

Americans have more timesaving devices and less time than
any other group of people in the world.[1]
                      —Duncan Caldwell

The High-Entropic Life in the United States

> The harder we try to control time, the more time con-
> trols us. *Why should that be?*[2]
> —Ralph Keyes

We are taught from childhood that time is a precious commodity. So when I first came to the United States, I quickly made some comparisons between life here and life in Alexandria, Egypt, where I grew up, to see whether I was going to have more time for myself. The differences were immediately noticeable. Here, the schools and businesses are open five days a week, while in Alexandria they are open six days. Because the weekend is twice as long, my first impression was that Americans must have lots of time for themselves.

A few days after my arrival in Fresno, California, my cousin took me to California State University at Fresno, where I was to begin my undergraduate studies in physics. When we arrived, I noticed the parking lots were full of cars, prompting me to comment that there must be quite a few professors and employees at the university. "No," said my cousin, "the great majority of these cars belong to the students." I translated this fact to mean that students here—and people in general—have more available time than their counterparts in Alexandria, because they do not have to wait for buses, tramways, or other means of transportation. They can go directly to their points of destination.

I also compared the supermarket system with the neighborhood stores that we had in Alexandria. Here, instead of going from one store to the other for meat, chicken, fish, bread, fruits, and vegetables, Americans can purchase everything from one large supermarket. Because of refrigerators and freezers, canned foods, and preservatives, Americans do not have to go every day to the market—another savings of time.

The widespread use of labor-saving devices like vacuum cleaners, dishwashers, telephones, washers, and dryers was also noticeable. And because there is no afternoon siesta, Americans commute to work only once, thus accruing more time for themselves. The situation was enviable. I was elated, thinking that I will have, like everybody else, plenty of time for myself.

As it turned out, the observations were accurate, but the conclusions were incorrect. As I gradually began to know people from this quiet community of Fresno, the most often-heard complaint was the "lack of available time" to get things done. I could not figure out the puzzle: How can people be surrounded with so much timesaving machinery, and yet be suffering from such an intense thirst for more available time? "It cannot be," I said, "it's just impossible."

Had I known the Laws of Thermodynamics then, I would not have been so perplexed, because thermodynamics compels us to look at the total thermodynamic equation before jumping to conclusions. In fact, the First and Second Laws of Thermodynamics explain nicely why people in the United States have less and less available time as they produce and attempt to use a wide spectrum of timesaving devices, which are not only increasing in numbers but also becoming increasingly complex and delicate.

We normally forget that it takes energy (work) to produce all these productivity-boosting machines. If work is required to build machines, it means that the users of cars, vacuum cleaners, microwaves, computers, washers, and dryers must also perform some work to pay for them. Consequently, more and more people are now working to pay for all the timesaving machines we all have in our possession.

"It was not supposed to be this way, of course," remarked the *Los Angeles Times*. The newspaper points out that in the 1950s and 1960s, when machines began replacing workers at an amazing rate, many economists, sociologists, and futurists predicted that the shift away from an economy dominated by the production of food, clothing, and other necessities of life would soon lead to less work and more leisure for everyone. "But it hasn't worked out."[3]

Time magazine also recalls "the silvery vision of a postindustrial age." In a cover story entitled "How America Has Run Out of

Time," the magazine remarked that "computers, satellites, robotics and other wizardries promised to make the American worker so much more efficient that income and GNP would rise while the workweek shrank." A testimony before a Senate subcommittee in 1967 indicated that by 1985 people could be working only 22 hours a week or 27 weeks a year, or could retire for good at age 38. This would leave plenty of leisure time for everyone to enjoy.[4]

The dream of working just 22 hours a week or retiring at 38 did not materialize because we live in a thermodynamic world, not a mechanistic one. There are many hidden thermodynamic variables. For example, in the case of automobiles, we cannot operate them in a vacuum. We need to build and maintain an intricate infrastructure of roads, highways, and bridges. Consequently, we have to work and pay for their expenses. Moreover, because automobiles can cause bodily harm and property damage, we have to work extra to pay for car insurance.

Automobiles emit many gaseous substances, which we call "pollution." Auto emissions change the composition of the air we breathe by adding unpleasant and often hazardous materials to the environment. The accumulated entropy creates a host of problems that demand more and more of our attention: the air pollution irritates our eyes and affects our lungs and general health, requiring additional medical attention.

The degradation of the quality of our environment necessitates the formation of watchdog bureaucracies—such as the Environmental Protection Agency—that are supported through additional work and taxes. These bureaucracies conduct studies, and write regulations that require our compliance. These regulatory laws call for research and development, usually resulting in the introduction of more gadgetry to "control pollution." All these activities require work and energy transformations, which in turn generate entropy and dissipate time.

Possessions Generate Entropy and Dissipate Time

If I keep a cow, that cow milks me.[5]
—Ralph Waldo Emerson

Today, Americans feel harried partly because of the demands on their time to purchase, use, store, maintain, replace, and protect all their labor-saving possessions. Many of today's high-tech machines do not even save time, especially after we factor in the setup and cleanup time. "We have a food processor that never comes out of the cupboard," states a physician. "The amount of time it takes to clean the thing after you chop two carrots just isn't worth it."[6]

Many people buy videocassette recorders with the intention of taping TV programs that are broadcast when they are busy, so they can watch them at their leisure. "After we waste several hours trying to install the device and hire someone else to do it, we are equipped to tape shows that are on when we're out," notes social commentator Ralph Keyes. "In order to choose them, we must read expanded television schedules in greater detail than we used to, including the wee morning hours our eyes used to skip over. (One woman I know has designed an elaborate weekly grid with which to plan her weekly taping.)"[7] Now that we have more than 100 channels of TV broadcasts, it is not a trivial task to go through all the programs to see which ones are worth the trouble of taping.

These machines make us feel that they are freeing us from the tyranny of synchronized life. We no longer have to watch the evening network news at 6:30 or 7:00 P.M. sharp. We can now program the machine to tape the news. We can go play tennis and then watch the news at our convenience when we come home.

What we forget, however, is that these machines generate additional activities (entropy), which in turn put more demands on our already hectic existence. Before the advent of VCRs, if we decided to play tennis during network news or situation comedies, we would do so and simply miss these programs. But now, every time we leave the house, we have to sit down and go through all the programs we might miss while we are out, then make decisions as to which ones are worth watching, and then program the VCR to record them. In the process, we have fallen behind in our activities. When we come home, we have all these programs to watch, in addition to the programs currently being broadcast—as well as everything else we like to do. "Home videotaping is just one among many techno-saviors which end up costing great gobs of time in the guise of saving some," remarks Keyes.[8]

"Americans are eating up their leisure time by overloading themselves with all kinds of gadgets, until they are worse off than they were before they had all these possessions."[9] That is the judgment of Ernest Dichter, whose pioneering work in marketing and motivational research has helped fuel the consumption spree by inducing consumers to acquire things they had not known they needed. More than half a century ago, British literary critic H. V. Routh observed: "Possessions might take more from the possessor than they gave."[10] Whether a possession takes more from us than it gives is in many instances a value judgment. What is not a value judgment, however, is the thermodynamic assertion that each possession generates entropy, which in turn dissipates time.

The transformation of recreation from simple pleasures—like reading, strolling, and visiting with friends—into complicated, capital-intensive, high-tech machinery, like recreational vehicles, takes its toll. "Not only do people who work additional hours to pay for such expensive playthings have less free time to enjoy them," observed the *Los Angeles Times*, "but essential maintenance chews up a lot more time than they bargained for."[11]

The servicing of all the machines in our possession dissipates a considerable amount of time. The Second Law brutally deteriorates them and breaks them down. The car is no exception. Although some items, like batteries and tires, may last longer than they used to, today's cars are more prone to breakdowns. To keep prices down and still maintain profits, carmakers have replaced durable materials with substitutes that are more fragile. Economics writer James Dale Davidson calls this phenomenon "The Quality Squeeze."[12]

In addition, today's cars have many more gadgets, which means there are more things that can break down. And thanks to techno-

logical advances, they have become amazingly complex. Car own-
ers can no longer tinker with simple screwdrivers and wrenches
under the hood and make repairs themselves.

Even the best-equipped dealers are having difficulties pinpoint-
ing and fixing certain car problems. One manufacturer's confes-
sion, that "with some kinds of electronic problems, it might be
required for the customer to come back 30 times," helped lead to a
"lemon law" in California.[13] Buyers of new cars are entitled to a re-
fund or a replacement car if the same defect is repaired four or more
times.

In recent years, many observers, noting the increase in the number of persons employed in the service sector, have drawn the conclusion that we are enjoying a general improvement in services.
Three decades ago, and before the explosion of high-tech products
that are nearly impossible to service, Swedish economist S. B. Linder
pointed out in *The Harried Leisure Class* that "the decisive factor for a
high standard of service is the service item rather than the total volume of services. A deterioration in the *quality* of service perhaps
provides a better description of existence in the rich countries than
the increase in *quantity* of service. We have good reason to speak of
'the decline of service in the service economy.'"[14]

Many of the conveniences we buy are "automatic," designed to
save us time and simplify our lives. But these cybernetic gadgets are
intricate, so they come with complicated operating manuals that
take time to read and understand. "Compared with a previous generation of medium-tech gadgets whose operations were basically
self-explanatory, the operation of electronic ones is anything but,"
observes Ralph Keyes. "Most require scriptural analysis of manuals
to reveal their secrets. What's worse, manuals are not only necessary to get modern conveniences up and running but to do even the
simplest maintenance."[15] And because high-tech devices have become so complex to operate, many manufacturers have set up, out
of necessity, toll-free telephone numbers to offer customers help and
reassurance.

In his first State of the Union message, President Nixon expressed bewilderment about why so many people were discontent,
despite the abundance of consumer goods. He put it this way:
"Never has a nation seemed to have had more and enjoy less."[16]

Actually, the Laws of Thermodynamics explain the phenomenon. For example, when we buy a car and drive it around, we are

elated. But when we have to work harder and harder, month after month, to pay for it, the enjoyment begins to diminish. When the amazing cybernetic gadgets we have in the car begin to deteriorate and break down, and we have to make many costly and time-consuming trips to the repair shops, the aggravations start to wear us down. When our eyes begin to burn and our lungs have to work overtime because of the pollutants in the air, when we are caught in a traffic jam at the least opportune moment, when gasoline prices shoot up unexpectedly, when our car insurance goes up because repair and other costs have risen, when our taxes are increased to pay for the roads and highways and to fix those cruel potholes that keep tearing apart our delicate possession—when all these things happen, and more, it is understandable why tension builds up within us. Those wonderful TV ads look much less attractive. We are no longer in the world of fantasy and simulation. We are in the world of thermodynamics, paying our dues.

Drowning in a Sea of Words

Complexity has a consequence that seems to draw very little comment. The explosion of data is heralded as leading to a new "information age." I suspect it may lead to a narrowing rather than a broadening of intellectual horizons. The Internet myth is riding high right now. Supposedly, it will make everybody with a computer into a universal genius.[17]
—Lindsey Grant

We are all elated when technology develops machines that increase our productivity. We feel we are going to produce more in less time and thus have more time for ourselves. We seldom analyze the situation in terms of the total thermodynamic equation. For example, computer technologists become ecstatic when they invent processes that increase printer speed. Computer salespeople are gratified when they convince their clients to throw away their older printers in favor of the latest, high-performance ones.

The same people, however, when they go to work and notice an increase in mail activity, complain they cannot keep up with what is going on. Their aggravation is heightened when they come home from a hectic day at work, only to find their mailboxes stuffed with junk mail. "The deluge is upon us," says *U.S. News & World Report.* "Driving the explosive growth in the direct-mail business is a vast data-collection and data-crunching network." Every American receives yearly, on average, 553 pieces of junk mail, a total of 4.5 million tons of material—nearly half of which goes directly into the garbage.[18] What's more, the volume of junk mail is expected to triple in the next decade.

Before word processors and computer networks hit the marketplace, futurist Alvin Toffler vividly described in *Future Shock* our utter frustration in coping with the blizzard of words, concepts, opinions, and theories that bombard us relentlessly. Scientific work is no exception. Here are some scientists' comments on information overload: "You can't possibly keep in touch with all you want to"; "I spend 25 percent to 50 percent of my working time trying to keep up with what's going on"; "I don't really know the answer unless we declare a moratorium on publications for ten years."[19]

The United States is devoting great efforts toward improving its productivity in generating information—that is, toward increasing the number of words that can be produced and disseminated per unit of time. Consequently, Americans are drowning in a sea of words. In 1981, John L. Kirkley, then editor of *Datamation*, probably expressed the feeling of many editors and readers alike when he remarked: "This probably sounds like heresy coming from a magazine editor, but there's just too damn much to read."[20]

When Kirkley made that statement, word processors, computers, and telecommunication networks were in their infancy. They are now sweeping through technological societies—especially the United States—and mesmerizing almost everybody along their way. Even Alvin Toffler, who reported humanity's immense difficulties in keeping up with the deluge of words and ideas, became fascinated with word processors when he started using them. "To learn how— and to speed up my own work—I bought a simple computer," confides Toffler in *The Third Wave*, "used it as a word processor, and wrote the latter half of this book on it. . . . After more than a year at the keyboard I am still amazed by its speed and power."[21] When he matched the slow typewriter with the speedy microprocessor, there was no contest.

The question remains, How do word processors help those who complained a decade earlier to Toffler that they could not keep up with what was going on? Are these amazing word processors a solution to the problems reported in *Future Shock*? Or are they machines that substantially aggravate an already bothersome situation?

Word processors are text editors that perform a variety of tasks, from simple deletion and correction of words to moving various paragraphs around. These capabilities definitely speed up the revision process, thus increasing the productivity of people who generate words. Word processors, however, do not detect or weed out ideas that are unworkable, irrelevant, insignificant, contradictory, redundant, meaningless, or contrary to the established Laws of Nature. They just execute commands that speed up the shuffling of words. They make it a breeze to delete, add, and change a word here and a sentence there, a paragraph here and an idea there, and to easily generate another piece of work—an article, a proposal, a contract, a piece of legislation, a book, or a scientific publication. They are particularly well suited to these activities.

The proliferation of word processors and computer networks has a profound effect on technological societies like the United States. By generating such massive amounts of words per unit of time, they depreciate the value of ideas and concepts, which become diffused in a turbulent sea of words. A given idea receives less and less attention. No matter how much time we spend in learning techniques to read faster, we still cannot keep up with the productivity gains in word generation. Consequently, we tend to confine ourselves to narrower and narrower interests, leading to shallower and shallower knowledge.

Word processors and other office automation products, like computer networks and workstations, voice messaging, facsimile, electronic mail, and satellite communication, are all touted as time-saving miracle tools of the information age. "There is no question that these tools, especially when all are available together, greatly improve the productivity of those who must write and distribute documents," wrote Peter J. Denning, past president of the Association for Computing Machinery. "But such tools may also increase the verbiage without increasing the number of ideas: they make it easy to combine an assortment of paragraphs from various files and call the result a document. Who will save the receivers from drowning in the rising tide of information so generated?" And as we install more and more of these entropy-producing machines, we will be increasing what Denning called "Electronic Junk."[22] Three decades ago, philosopher and social critic Lewis Mumford coined the term "Electronic Entropy."[23]

American workers—including information technologists themselves—are being submerged in piles of data produced by information technology. Those who invent the productivity tools are not immune from data overload. For example, the director of product management for Exchange, a communication software, reported accumulating 2,700 e-mail, voicemails, faxes, and other messages in his mailbox in just two weeks—and that was in 1995.[24] Thanks to relentless advances in applied physics, computers and communication devices are becoming increasingly faster and interconnected, allowing more of us to produce, send, and receive bits of information at increasingly faster rates and bigger chunks. Not surprisingly, the word "infoglut" has become part of our vocabulary.

In his book *In Praise of Idleness*, philosopher Bertrand Russell regretted that humankind has chosen a path in which every productivity gain has been used to increase the number of material goods, all of which make their demands on us. These consumer goods devour valuable time that could have been used to cultivate our mind. He deplored our learning to make twice as many pins in a given time, instead of making a given quantity of pins in half the time.[25]

As it turns out, people who produce material for the cultivation of the mind—ideas, symbols, theories, and opinions—are in the same trap as product manufacturers, as evidenced by the massive explosion of words around us. This should not come as a surprise. Every productivity gain is achieved through some machinery, which in turn absorbs energy from the environment while increasing the entropy of the thermodynamic system. The Laws of Thermodynamics tell us that whichever way we decide to pay for this increase in entropy, we cannot do it without performing additional work—that is, without generating some more entropy. And the process feeds on itself. This is what we are witnessing in the information age—an explosion in "Information Entropy."

More Choices but Less Time

Because modern computers and communication systems can store, manipulate, and distribute massive amounts of data instantly, they have been used to provide consumers with a dizzying array of options. Take the airline industry. Aided by government deregulation of air fares and supported by sophisticated computer networks, airlines have created an unbelievable variety of fares—exhausting sales clerks and travel agents while confusing travelers. The daily changes alone are overwhelming.[26]

What used to be a simple travel decision now requires a cost-benefit analysis to choose between the high-priced fares that have virtually no restrictions and the low-priced ones that are loaded with instructions. While it is a simple task to enter verbiage about bargain fares one time into the computer, and then to broadcast it instantaneously to all travel agents connected to the computer network, the generated entropy devours much time and effort from agents and travelers alike.

Choices, alternatives, varieties, and options absorb considerable time from American society. Shopping, for example, has become much more time-consuming. Since 1950, the typical supermarket has increased its stock from an average of 2,500 to about 13,000 items.[27] Studies have found that shoppers give an item of interest an average of four seconds of consideration.[28] Just walking through all the aisles can more than equal the effort of strolling to the neighborhood mom-and-pop stores of the old days. Waiting for checkout can also gobble up time, especially if coupon-redeemers and check-cashers are ahead of us.

Furthermore, if we attempt to read and understand all the health and nutritional information printed on the packages in order to make an informed buying decision, then shopping can truly become a time-consuming affair. And should we add environmental issues to what we buy, shopping can turn into a major project. Just figuring out what the manufacturer means exactly by "biodegradable" or "photodegradable" may require the effort of a master's thesis.

Americans have less and less time to inform themselves fully about an increasing number of complicated products and services—from health care to life insurance. Consider money management. Americans with a little extra cash used to open a passbook savings account and go about their business. It is no longer that simple. They now have to choose among a bewildering array of rapidly changing alternatives, including certificates of deposit, money market funds, and retirement accounts, each bearing different interest rates, terms, and maturity dates. There are now three types of Individual Retirement Accounts; six plans for small businesses or the self-employed; and several employee plans. All have different rules. We cannot necessarily move from one type of plan to another. In a *Newsweek* article, "The Virtues of Simplicity," columnist Jane Bryant Quinn writes: "At retirement, the withdrawal rules can send you to bed with a migraine. For that matter, writing about them does, too."[29]

Business executives are among the most harried people in the United States. Their minds are seldom at rest. A 1998 survey of the vacation habits of business executives found that the majority of them have called their office while on vacation, have checked work phone messages, have received a call or e-mail from the office, or have spent time doing office work.[30] The explosion of data base centers and communication systems has increased the pressure on business executives (along with everybody else) to research topics more thoroughly. Once businesses are connected to data base centers, there is nothing they will not try to find out. Moreover, the transportation revolution has made it easy for executives to hop onto jets and attend meetings.

The high-entropic life is taking its toll on Americans. "Harried, hurried and haunted, they rush through their lives under the tyranny of the clock in an Age of Haste," remarked the *Los Angeles Times*.[31] In an environment of uncertainty and high-cost of living, more Americans are now working in order to survive. No one imagined that the middle class would be squeezed; the prices of houses have shot up, inflation has eroded paychecks, wages have remained stagnant, and medical and tuition costs have skyrocketed. Thus it now takes two paychecks to support what many thought was a middle-class life.

For the bottom third to half of the work force, stagnating wages have increased pressure on both parents to work. Wives in the poorer half of the U.S. households typically work 40 percent more hours today than two decades ago. "At the bottom," states economist Sheldon Danziger of the University of Michigan, two sources of income are "the difference between making ends meet and not making ends meet."[32] Despite millions of families with a second earner in the workplace, "somewhere between a quarter and 30 percent of households live paycheck to paycheck."[33] And 14 million American children grow up below the poverty line.[34]

According to *U.S. News & World Report*, the cost of bringing a typical child into this world and raising him or her through college "requires a 22-year investment of just over $1.45 million" for a middle-income family. "That's a pretty steep price tag in a country where the median income for families with children is just $41,000."[35]

Women are especially feeling the high-entropic life. Keeping a home and raising children is a full-time job in itself. But in the United States, close to 60 percent of wives with children age one or younger are also working outside the home.[36] As the cost of domestic help has soared, far fewer women have household help than their mothers and grandmothers did. In addition, the breakdown of neighborhoods and the dispersal of families have removed much of the help with children and chores that mothers used to rely on from relatives and neighbors. In 1990, the majority of American infants and toddlers were in the care of someone other than their parents.[37]

In the United States, the number of divorces has tripled since 1960, while the number of single-parent families has more than quadrupled.[38] With the majority of mothers of minor children now working, fragile and impressionable young minds are being farmed out to day-care centers. Parents who are too pressed for time have resorted to giving their kids concentrated dosages of focused attention, which they call "quality time." As professor of sociology Arlie Russell Hochschild points out in *The Time Bind*, the concept of quality time was unknown to our ancestors. Quality time "has become a powerful symbol of the struggle against the growing pressures on time at home. It reflects the extent to which modern parents feel the flow of time running against them."[39] This is happening at a time when parents are surrounded with more timesaving conveniences than ever before.

Children, too, are feeling the weight of this hectic, high-entropic family life. They understand that they are being cheated out of childhood; eight-year-olds are taking care of three-year-olds. Kids are sensing that adults are ignoring them. Children are also being afflicted with depression in great numbers. The American Academy of Child and Adolescent Psychiatry estimates that "3.4 million children suffer from depression, but that fewer than half receive treatment."[40] In a cover story on the "Melancholy Nation," *U.S. News & World Report* notes that "many who study depression say that we are entering an 'Age of Melancholy,' where people are getting depressed at younger and younger ages, with episodes that are severe and frequent."[41]

Psychologist David Elkind believes children are growing up too fast too soon, feeling pressed for time and worse off emotionally.[42] In *Ties That Stress: The New Family Imbalance*, he writes, "children today are under much greater stresses than were children a generation or two ago, in part because the world is a more dangerous and complicated place to grow up in, and in part because their need to be protected, nurtured, and guided has been neglected." He points out that "young people who are stressed often do what adults do: they engage in actions that are destructive to themselves, to others, or to both. The consequences of these self/other punishing practices are increasingly evident, and have been given a new name, the 'new morbidity.'"[43]

As their parents struggle to cope with divorce, single parenthood, dual careers, eroding paychecks, poverty, addiction to a speeded-up schedule, and an uncertain economic future, many of the nation's children are paying the price—in ways ranging from simple neglect at home to outright abuse.[44] *Los Angeles Times* contributing editor Andrew Vachss characterizes today's children as "Our Endangered Species."[45]

Are We Freeing Ourselves from Machines, at Last?

Man rushes first to be saved by technology, and then to be saved from it. We Americans are front runners in both races.[46]
—Gerald Sykes

In America and other advanced cultures, belief in technology has become a religious faith. . . . We have become its servants rather than its masters.[47]
—Mary Eleanor Clark, *Biologist*

Since the industrial revolution, machines have become an integral part of modern society. As machines increased in number, and became faster and more complex, our lives became more mechanized and more hectic. Each technology promised to make our lives easier and give us free time, but instead made our lives more frenzied and more attached to machines. Some technology enthusiasts believe, however, that machines have finally become so sophisticated that they are actually freeing us from their tyrannical pace. In *The Third Wave*, Alvin Toffler explains the phenomenon this way: "The reason is that the Third Wave [computer, communication, and information age], as it sweeps in, carries with it a completely different sense of time. If the Second Wave [industrial age] tied life to the tempo of the machine, the Third Wave challenges this mechanical synchronization, alters our most basic social rhythms, and in so doing frees us from the machine."[48]

How are we freeing ourselves from machines? Are we gradually eliminating them from our environment? Not at all. On the contrary, we are rapidly introducing ever-faster machines. If the Third Wave is freeing the secretary from the tempo of the typewriter, it is doing so by ushering in the word processor, which is not only a faster tempo input/output machine but also a computer that can monitor the secretary's input/output performance. If the engineers of the industrial age were driven by the slide rule, the engineers of the Third Wave are driven by the computer, a faster and more complex machine.

When workers of the industrial age came home at the end of the day, they were truly free—physically disconnected—from the machines of the factory. But in today's environment, the advanced machines of the Third Wave—beepers, voice-messaging units, fax machines, computers, and communication systems—are following workers (and humanity) everywhere: at the office, at home, in the car, in the streets, and even in an airplane.

We are now surrounded by all kinds of machines, constantly prompting us to do one thing or another. We have telephones in the car, computers in the den, and the humming fax machine has eliminated that once peaceful pose between completing a document and delivering it. The fax machine is one of the latest inventions of the information age that synchronizes the workers to its relentless pace. It has demolished any sense of patience or grace that existed.[49]

On the other hand, some technology advocates contend that the workers of the information age are being decoupled from machines. Again, Toffler:

> Today, machine synchronization has reached such exquisitely high levels, and the pace of even the fastest human workers is so ridiculously slow by comparison, that full advantage of the technology can be derived not by coupling workers to the machine but only by decoupling them from it.
>
> Put differently, during Second Wave civilization, machine synchronization shackled the human to the machine's capabilities and imprisoned all of social life in a common frame. It did so in capitalist and socialist societies alike. Now, as machine synchronization grows more precise, humans, instead of being imprisoned, are progressively freed.[50]

If we are decoupling and freeing ourselves from machines, we are doing it most precariously, by bringing in ever-more-complex machines and connecting ourselves to them in record numbers.

Today, more workers than ever before are coupled to more machines. Before, in the industrial age, many white-collar workers—

bankers, librarians, tellers, lawyers, insurance agents, teachers—
were hardly connected to any machine. They are now connected to
computers, modems, communication systems, data banks, electronic mail, and fax machines. In fact, high-tech machines have attained such a high level of sophistication that they can be set up to
monitor practically anyone's work performance and keep tabs on
workers' activities.

Computer monitoring techniques are being used in hotels, supermarkets, restaurants, airlines, offices, and data processing centers. Nurses are monitored through a box on their belt that tracks
the amount of time used for each procedure with a patient. Truck
drivers have on-board computers that track how many stops the
driver made and where. Hotel staff punch a code into the telephone
when entering and leaving a room, thus providing a detailed log of
their cleaning activities for the entire day. Data processors are being
monitored by computer networking software; a data processor's
computer screen periodically flashes the message: "You're not
working as fast as the person next to you."[51] In 1987, the U.S.
Congressional Office of Technology Assessment published a report
acknowledging that "the present extent of computer-based monitoring is only a preview of growing technological capabilities for
monitoring, surveillance and worker testing on the job."[52] A 1998 a
survey of 1,085 corporations conducted by the American Management Association revealed that "more than 40 percent engaged in
some kind of intrusive employee monitoring. Such monitoring includes checking of e-mail, voice mail and telephone conversations;
recording of computer keystrokes; and video recording of job performance."[53]

That machines are monitoring our performance (but not necessarily the quality of our work) is evident in the fast-food business. At
one such restaurant, you get a dollar off if your order is not served in
60 seconds. So the computer keeps track of the time elapsed between
the placement of the order and its delivery. As one employee explained, "Every 15 minutes, it prints up a new average for us," telling them how they are doing.[54]

Today's jobs are "faster and more punishingly repetitive."[55] In
the supermarket industry, for instance, the installation of timesaving laser scanners and computerized cash registers on the checkout
lines has been a catalyst to an epidemic of worker injuries, known
collectively as cumulative trauma disorders (CTD). These painful

ailments include tendinitis and carpal tunnel syndrome, and are linked to repetitive actions such as the wrist motion a checker must make to pass products over a scanner.[56] In the United States, CTD has become a frequently reported occupational illness, and has been dubbed the "industrial disease of the Information Age."[57]

High technology has given many traditional jobs a precarious twist. For example, secretaries two decades ago rarely developed carpal tunnel syndrome, an ailment concentrated in the carpal bones of the hand and wrist, because typewriters required frequent paper changes and manual adjustments. By contrast, computers enable typists to move much faster, offering no break from the keyboard. Often carpal tunnel syndrome sufferers, whether they are office workers, supermarket checkers, or meatpackers, have trouble lifting a cup of coffee. In the worst cases, the pain and numbness can lead to permanent disability.

In this information age, whichever way we turn, we are being linked—for whatever purpose—to a machine of one kind or another. In almost all sectors of the economy, businesses have set up computers to provide customer service. Consequently, when we call a bank, a telephone company, a computer manufacturer, or the Internal Revenue Service, we are greeted by an automated telephone answering system, which often sends us into a loop. These answering machines were installed in droves to save us time and give us better care.

Ironically, customer-service satisfaction has dropped overall since 1994, as measured by the American Satisfaction Index compiled by the University of Michigan. "There are many causes, but one that is widely recognized is technology itself," remarks the *New York Times*, adding: "Phone systems and Web sites that are supposed to help can turn into infuriating mazes."[58] In recent years, complaints to the Better Business Bureau have risen sharply. "It's a fascinating paradox," notes Laurel Pallock, an investigator who runs the complaint mediation program in the San Francisco district attorney's office. "Here we are in this predicament because of technology, the very thing that is supposed to keep us from being in this predicament."[59]

Chapter 6

The Agricultural-Industrial Complex

Increased use of pesticides, herbicides and fertilizers is akin to taking drugs. We get a high but it takes more and more of the stuff to produce the effect. Sooner or later the increased dose becomes toxic or we run out of money to support our habit.[1]

—John E. Thompson, *Agriculturist*

Modern Agriculture and the Second Law

At a time in human history when food supplies seem to be plentiful, a book appears with the ominous title *Tough Choices: Facing the Challenge of Food Scarcity.*[2] Is this alarm unwarranted? Not at all. Not when world population is growing at about 80 million a year. Not when industrialization and economic growth are forcing the conversion of cropland to nonfarm uses. Not when the fertility of our farmland is deteriorating faster than Nature requires to replenish it. Not when the Second Law of Thermodynamics is warning us to behave or pay the dire consequences.

Humans can increase food supply by two basic means: by bringing more land under cultivation or by squeezing more food out of existing farmland. Throughout most of history, our efforts have been directed toward the first method. In addition, we have developed technologies to raise the productivity of land. Recently, because of the decreasing availability of new lands for expansion, we have turned more and more to high-intensity agriculture, placing ourselves on a direct collision course with the Laws of Nature.

The high-intensity agricultural techniques we have been practicing in recent years, especially since World War II, are beginning to take their toll both ecologically and economically. In addition to stripping the land of its fertility, the profligate methods involve altering the biosphere's cycles of energy, water, nitrogen, and minerals. High-intensity agriculture depends heavily on four technologies: mechanization, irrigation, fertilization, and chemical control of weeds and insects through herbicides, insecticides, and fungicides. All four technologies have helped raise yields per hectare, but they have also perturbed the cycles of the biosphere and have contributed to substantial increases in the entropy of our thermodynamic system.

For a long, long time, humans harnessed powerful animals to augment muscle power. At least as early as 3000 B.C., farmers in the Middle East learned to harness draft animals to till the soil. People also returned animal waste to the soil, which in turn helped the soil-rebuilding process. The invention of the internal combustion engine and the tractor drastically changed the manner in which we interacted with Nature. These technological advances made it possible to substitute petroleum for the oats, corn, and hay grown as feed or fuel for draft animals.

In the transition from horses to tractors, humans replaced an essentially renewable source of energy with a nonrenewable one. More important, in today's highly mechanized agriculture the input of fossil fuel energy per hectare is often substantially greater than the energy yield contained in the food produced.[3] Although it appears that high-intensity agriculture is extremely efficient, the discipline of thermodynamics gives this technique failing marks because the system often operates, energywise, with a deficit. "The most efficient agriculture is not the mechanized agriculture of the industrial world," writes geologist Earl Cook in *Man, Energy, Society*, "but that which minimizes energy input and maximizes energy output, without depleting the resources of the soil."[4]

The energy costs of mechanized agriculture are high because the tools and machinery are large and take a substantial amount of energy to manufacture, operate, and maintain. Tractors and farm equipment are capital-intensive. In addition to consuming lots of energy, they also deteriorate with use. With the passage of time, it becomes costly to maintain such systems.

The second technology that modern agriculture depends on is irrigation, which farmers began to use at least 6,000 years ago. By bringing into cultivation vast areas that would otherwise be unusable or marginally productive, irrigation has played a major role in increasing food production. Water, once thought inexhaustible, is the most limiting natural resource for crop production in semiarid agricultural areas. Water scarcity has become "the single biggest threat to global food production," notes Sandra Postel of the Global Water Policy Project.[5] Serious water problems now exist in every continent and are spreading rapidly. More than a billion people live in areas where there is not enough water to meet their modest food and material needs.[6]

When rainfall plus the water already stored in the soil is insufficient to meet crop needs, the inescapable consequence is a reduction in crop yields in proportion to the water deficit. The idea behind the installation of an irrigation system is to reduce or eliminate this water shortage. However, full irrigation in many semiarid areas is depleting groundwater supplies, thus lowering the water table. Consequently, the vertical distance that water must be lifted is continually increasing.

In many areas, farmers are pumping water from aquifers much faster than Nature has time to replace it. Every day U.S. farmers and ranchers extract 20 billion more gallons of water from the ground than are replenished by rain.[7] About a fifth of the nation's irrigated area is watered by overpumping, an unsustainable activity.[8] The situation is similar in other countries: In India, water tables are dropping in many states, including the Punjab—the country's breadbasket. Much of China's northern region is water-deficient, satisfying part of its water needs by overpumping aquifers. In Beijing, the water table has fallen from 5 meters below ground level in 1950 to more than 50 meters.[9] The gradual depletion of underground waters is inflicting economic hardships on farmers. As water tables decline, the cost of pumping water rises due to the capital required to deepen the wells and to the higher cost of energy to operate the pumps.

There is no question that some of the highest crop yields are produced by irrigation. However, like many of our interventions in natural processes, reshaping the hydrologic cycle has had undesirable side effects. One of them is the raising of water table levels by the diversion of river water onto the land. After a time, the percolation of irrigation water into the land and the accumulation of this water underground may gradually pull the water table up until it is too close to the surface. When this happens, the growth of plant roots is inhibited through waterlogging and the surface soil becomes salty as water evaporates through it, leaving a deposit of salts in the top few inches. At least 10 percent of the world's irrigated area is now suffering from waterlogging and salinity.[10] Roughly two thirds of the world's irrigated area needs some form of upgrading to remain in good working condition.[11]

"Although large-scale irrigation has buttressed the world against famine and helped eliminate some pockets of chronic hunger," points out Sandra Postel in *Last Oasis: Facing Water Scarcity*, "it

often has not served other important development goals, such as reducing poverty, promoting equity, protecting natural systems, and
improving human health. . . . Many people are beginning to ask, Irrigation for whom, and at what social and environmental cost?"[12]

Fertilizers are the third technology that humans have introduced to increase food production. The high crop yields that American farmers have achieved have been made possible by intensive
application of chemical fertilizers. We owe the foundation of this
development to the German chemist Baron Justus von Liebig, who
set the specific requirements for nitrogen, potassium, phosphorous,
and other nutrients used in plant growth.

Chemical fertilizers remained in the background until the start
of the last century. The pressure of population growth, coupled
with the disappearance of new frontiers, eventually compelled
farmers to turn to fertilizers to boost food production. Japan, the
Netherlands, Denmark, and Sweden were among the first countries to use these inputs. The United States, richly endowed with
vast farmlands, remained immune from heavy fertilizer usage until the 1940s. Since then, the United States has joined the club of
heavy fertilizer users. In return, yields per hectare have shown remarkable gains.

However, the beneficial effects of chemical fertilizers have deteriorated through time. The initial, easy production gains through
intensive fertilizer use are largely behind us. Fertilizers have simply lost their potency. As application rates increased, diminishing
returns eventually set in. Consequently, fertilizer use has plateaued
in many countries—including the United States, Japan, and most of
Western Europe.[13] "The old formula of combining more and more
fertilizer with ever higher yielding varieties to expand the grain
harvest is no longer working very well," points out agricultural expert Lester R. Brown, chairman of Worldwatch Institute.[14]

A cornfield is an effective way to trap solar energy (radiation)
and transform it into chemical energy (food). Modern agriculture,
however, while endeavoring to squeeze higher yields from the
land, has become a major consumer of energy through its heavy
use of fertilizers, which are derived mainly from fossil fuels. We are
using fossil fuels and chemicals to give a boost of specious life to
dying soil. "The effect of this is to make it die faster," writes nature and technology commentator Evan Eisenberg in *The Ecology of
Eden*. "Manure is something the soil population can use; chemical

fertilizer is an attempt to go over its head. The crops get plenty of nitrogen, phosphorous, and potassium, but next to nothing of the countless trace elements needed for plant and human metabolism."[15]

We have benefited from the use of fertilizers, but the benefits have not been unalloyed. The great increases in food production (which are now leveling off) have come about at the expense of massive increases in entropy. The runoff of chemical fertilizers into rivers, lakes, and subterranean waters has created some pernicious problems for humanity and other living organisms. One is the chemical pollution of our drinking water. Excessive nitrate levels in water can cause a medical disorder called methemoglobinemia, which has the effect of reducing the oxygen-carrying capacity of the blood. This physiological disorder can be particularly dangerous to children under five. The Organisation for Economic Co-operation and Development lists nitrate pollution as one of the most serious water quality problems in Europe and North America.[16]

The second hazard of fertilizer runoff, and the much greater one due to its profound effects on the environment, comes from the phenomenon called eutrophication. When inorganic nitrates and phosphates are discharged into rivers and other sources of fresh water, they provide a fertile medium for the growth of algae. The resulting massive growth of algae depletes the water of its oxygen supply and gradually kills off the fish. Eventually, the overfertilization (eutrophication) of a lake or river brings about its death as a source of fresh water. This environmental problem is apparent in the Gulf of Mexico, the Mississippi River's terminus, where a "dead zone" the size of New Jersey forms every summer. The once productive area has now far less fish and shrimp because of lack of oxygen due to decomposing algae.[17]

Chemical Control of Insects

While the use of fertilizers in many countries has reached a plateau, the application of pesticides—the fourth technology supporting high-intensity agriculture—has not leveled off. There are now more than 1,600 pesticides, and their worldwide use is still increasing.[18] "Extensive regulations appear to protect us from the most toxic chemicals but lull us into a false sense of complacency that our use of chemical pesticides has diminished," writes biologist Mark L. Winston in *Nature Wars: People vs. Pests*. "Not So. The extent and impact of our current dependence on pesticides for both agricultural and nonagricultural purposes is staggering."[19] In 1993, 4.5 billion pounds of active pesticide ingredients were used worldwide, and in the United States the figure was 1.1 billion pounds—about 4 pounds per person.[20]

Insects have plagued humankind since antiquity. As a hunter and gatherer, however, humans grew no crops to be raided. The situation changed with the passage of time. Populations grew, and animals alone could no longer support us. We became increasingly dependent on agricultural products for survival. Crop protection from insects became a serious problem. As we developed an understanding of the role of insects in the transmission of diseases, we began exploring and developing insect control mechanisms. Finally, as we became industrialized, we expanded our insect control efforts to include species that attack gardens, forests, dwellings, clothing, pets, parks, and livestock.

Despite our efforts to combat a widening array of species, until recently insect control remained a low-level activity. We simply accepted the losses inflicted by insects as a natural consequence of doing business with Nature. Then, in a flash, came what seemed an apparent miracle: DDT. This amazing chemical, which had been sitting on a shelf for decades, was found to be the most powerful insecticide ever tested. Having discovered the ultimate weapon, which would exterminate an unlimited spectrum of unwanted insects, modern humans went for the kill.

As an insect killer, DDT worked with deadly efficiency. It was found to be widely toxic, long-lasting, and inexpensive. It seemed to have all the characteristics of an ideal weapon, a miraculous product of technological know-how. In 1948, Paul Hermann Müller of Switzerland received a Nobel prize for discovering DDT. Ironically, today the use of DDT is banned by law in many countries, including the United States, because of its devastating effects on our environment and human beings. "This illustrates how little man knows about the effects of his intervening in the biosphere," observes Lester R. Brown. "Up to now he has been using the biosphere as a laboratory, sometimes with unhappy results."[21]

Today, despite all of DDT's insect-killing powers and the subsequent development of scores of powerful poisons, the bugs are doing better than ever. There are now more insect species of pest status than ever before. Insecticide costs have risen, and the insecticides' impact on the environment is at historic levels.

We have declared war on insects. Sadly enough, however, the lowly insects are humiliating the highly evolved and the highly advanced *Homo sapiens*. Insects are formidable foes. To survive, they use their great assets of diversity, adaptability, plasticity, and prolificity. Their successful formula has worked consistently for some 350 million years, and is now working with astounding efficiency in their combat with human beings.

There are at least a million insect species, but only a few thousand have attained pest status. Many species are harmless and most are probably beneficial, acting as pollinators, scavengers, and natural enemies of pests. Many insect species have the potential to become pests, but they fail because Nature keeps them in balance. Among the biological factors that infringe upon them, two kinds of natural enemies—the predators and the parasites—play a significant role. Without them, our rivalry with the insects would have been ferocious indeed.

Modern insecticides are changing the delicate balance that existed in Nature for millennia. Chemicals kill a broad spectrum of animals and insects. When we apply an insecticide to a crop, we kill not only the target pests but also other species in the insect family, including the natural enemies that restrain harmful species. Often, the natural enemies of pests suffer inordinately from the chemicals because they are generally less robust than the pest species. Moreover, as insecticides kill the pest species, there is no more food left for their predators. So the enemies of pests are either forced to starve or leave the fields.

Any surviving pests, free from significant natural enemies, now experience a population boom. "Such post-spraying pest explosions are often double-barreled," observes professor of entomology Robert van den Bosch, "in that they involve not only the resurgence of target pests but also the eruption of previously minor species, which had been fully suppressed by natural enemies." He describes what happens next:

> The frequent outcome is a raging multiple-pest outbreak, more damaging than that for which the original pest-control measure was undertaken. Predictably, the grower or other insecticide user, in order to salvage his cotton, fir trees, rosebuds, or whatever, re-applies insecticides, and when this triggers still another multipest outbreak, he sprays again. This is the genesis of the insecticide treadmill, an addictive process that is magnified and prolonged by genetic selection for insecticide resistance in the repeatedly treated pests.[22]

The development of resistance to stress in populations of organisms, such as that brought about by the spraying of toxic chemicals, is a normal evolutionary process. Insects have demonstrated an extraordinary ability to adapt to sudden and harsh changes in climate and habitat. They are endowed with remarkable genetic plasticity, the main reason they have lasted so many millions of years. They are accustomed to facing recurring adversities. They now have one more adversity to deal with—insecticides—and they are easily meeting the enemy.

As insecticides are sprayed, most bugs die, but the strong ones with traits of survival hang on, and they come to dominate the population. Due to their prolificity, they quickly multiply into a huge insect population with a built-in immunity to the insecticide.

Not surprisingly, insect-related costs and losses keep going up. After four decades of insecticide bombardment, the share of crop yields lost to insects has roughly doubled, despite more than a 10-fold increase in the amount and toxicity of insecticide used.[23] Modern insecticide technology is taking farmers slowly to the poorhouse.

There are now more than 500 insect and mite species that have developed resistance to pesticides, in addition to 150 plant pathogen species and 273 weed species.[24] With pesticide resistance plugged into the formula, we are caught in a dangerous pesticide treadmill, which is showing no sign of deceleration. We keep increasing the pesticide dosage, at the same time hoping that it will never reach a poisonous level so high that it inflicts greater damage upon humans and the environment than upon insects.

The use of pesticides is not confined to agricultural fields. As environmentalists Theo Colborn, Dianne Dumanoski, and John Peterson Myers remind us in *Our Stolen Future*, pesticides are "here, there, and everywhere," in parks, schools, restaurants, supermarkets, homes, and gardens.[25] The authors alerted the scientific community to the possibility that pesticides and other synthetic chemicals could be interfering with the reproductive process in some species of birds, fish, and amphibians, as well as with the human hormone system. In humans, these so-called endocrine disrupters are suspected of derailing the sensitive developmental processes in the unborn and the young, causing a multitude of later problems— from aberrant sexual development, to cognitive deficits in children, to lower sperm counts in adults.[26]

The impact of pesticides on humans and the environment has reached a noticeable level. According to a 1989 World Health Organization estimate, there are about 1 million human poisonings a year worldwide, with 20,000 deaths, mainly in developing countries, where regulations are loose.[27] Constant exposure to pesticides has been linked to such medical disorders as infertility, immune dysfunction, and various forms of cancer and birth defects. In 1993, professor of agricultural sciences David Pimentel of Cornell University estimated the annual cost of human pesticide poisonings and pesticide-related illness at about $800 million in the United States.[28]

There are other costs associated with pesticides, such as losses from livestock fatalities and destruction of pesticide-contaminated milk, meat, and eggs. Decontamination is even a bigger item. In the United States, the monitoring and cleanup of groundwater polluted by pesticides costs about $2 billion a year.[29] Pesticides also have an impact on nonhuman, nontarget species. Millions of birds and fish are killed by pesticides each year in the United States. We should not be surprised by this. After all, pesticides are developed to have a biological effect. In Latin, *cide* means kill. Pesticides are doing their job creating biological havoc.

Moreover, some of those pesticides, like DDT, remain in the environment for a long time. "Pesticides are ubiquitous in virtually all habitats, even far from sprayed areas decades after spraying, and have negative impact on everything from soil-dwelling microorganisms to plants to beneficial insects, in addition to vertebrates," writes Mark Winston. "The classic cases of insects developing pesti-

cide resistance and birds and fish accumulating pesticides and dying at potentially extinctive rates are just two examples of the many
instances in which our technological mastery over pests has in the
end failed us, as well as the organisms and ecosystems we claim to
steward."[30]

Soil Erosion and Degradation of the Environment

*All terrestrial life ultimately depends on soil and water. So
commonplace and seemingly abundant are these elements
that we tend to treat them contemptuously.*[31]
—Daniel J. Hillel

The deterioration of the quality of the environment and the erosion
of topsoil are the inescapable consequences of high-intensity agriculture. Soil erosion "is a chronic, slow process without the urgency of an environmental crisis. Yet, soil is the basis for our very
existence," points out Priscilla Grew, the former director of the Department of Conservation for the state of California. "Where soil is
lost, civilization often goes with it."[32] The Law of Increasing Entropy is acting on our topsoil in much the same way as it acts everywhere: persistently and cumulatively.

Soil is lost to erosion every year, but it is also continually being
formed. Under normal conditions, soil may form at the average rate
of 1 ton per hectare per year. Asia, Africa, and South America have
the highest rate of soil erosion, at 30–40 times the rate of soil formation. The United States and Europe have the lowest rates, still 17
times the rate of formation. In the last two centuries, an estimated
30 percent of U.S. farmland has been abandoned because of soil erosion, salination, and waterlogging. About 90 percent of U.S. cropland is losing soil above the sustainable rate.[33]

Topsoil deterioration is nowhere more dramatic than in the rich
but fragile land of the western Tennessee counties bordering the

Mississippi River. The destruction there is highly visible. About 3.5 hectares worth of topsoil floats past Memphis every hour. The Mississippi River carries away millions of tons of valuable topsoil from farms in the middle of America—soil that is now gone for good.[34] We know we cannot do that forever.

Iowa, which provides 20 percent of the entire U.S. corn crop and 15 percent of the soybean harvest, is also experiencing serious soil degradation. About a century and a half ago, when Iowa was first plowed and planted, it was covered with rich black topsoil to an average depth of 16 inches; now it's down to 8 inches.[35] Iowa continues to lose topsoil at the rate of 30 tons per hectare per year. Similarly, about 40 percent of the rich Palouse soils of the northwest United States have been lost.[36] The use of heavy machinery in agriculture is damaging the entire soil ecosystem.

Nature has an amazing way of collecting taxes, which is through increases in the entropy—disorder—of the thermodynamic system. The more intense means we use to pluck food from Nature, the more we increase the entropy of our environment.

Topsoil degradation leads to other environmental disorders. As topsoil erodes, the land loses its natural capacity to hold and absorb water. Whenever we observe severe flooding and swamping conditions, we are only seeing the effects of the real problem—the deterioration of soil structure. When soil erodes, it does not simply move from one field to another. Some does, but most soil ends up in roadside ditches, streams, reservoirs, rivers, and lakes. Not surprisingly, agriculture has become a major polluter of waterways throughout the world. In several European countries, drinking water is contaminated with fertilizer runoff.[37] Farmers have been attempting to mask the negative effects of soil erosion by adding fertilizers which, together with pesticides, have ended up in waterways.

Modern high-intensity farming has been hailed as the Green Revolution, whose major architect is the American agronomist and Nobel laureate Norman Borlaug. While it is true that the Green Revolution has provided much food to the world, the fact remains that it has done so through massive increases in the disorder of the thermodynamic system. High-intensity farming has been a boon to the "Agricultural-Industrial Complex," which provides the farmer with machinery, pesticides, fertilizers, and energy. But the high-entropic methods are not only costly but create havoc in the environment.

In *The Heat Is On*, reporter Ross Gelbspan writes that Borlaug "intended the Green Revolution to be a short-term effort by which poorer countries could develop modern, sustainable growing practices." Unfortunately, this has not been the case. As David Pimentel and physics Nobel laureate Henry Kendall of MIT point out, "the Green Revolution has been implemented in a manner that has not proved environmentally sustainable. The technology has enhanced soil erosion, polluted groundwater and surface water resources, and increased pesticide use has caused serious public health and environmental problems."[38]

Humanity is beginning to feel the consequences of the Laws of Thermodynamics, which are asserting themselves with great authority. "Crop and animal yields can be increased significantly," remarks chemist G. Tyler Miller, Jr., "but many agricultural optimists have forgotten or do not understand the environmental penalty extracted by the second law."[39] In agriculture, as in all areas of human endeavor, it is crucial that we pay attention to the Laws of Thermodynamics, especially the Second Law.

Chapter 7

What Does the Second Law Really Say?

*I found the meaning of the Second Law of Thermodynamics
in the principle that in every natural process the sum of the
entropies of all bodies involved in the process increases.*[1]
—Max Planck

The Availability of Energy and Natural Resources Revisited

Man, as a conscious and constant, single, natural force, seems to have no function except that of dissipating or degrading energy. . . . As an energy he has but one dominant function:—that of accelerating the operation of the second law of thermodynamics.[2]
—Henry Adams

From its inception, the Second Law of Thermodynamics has been depicted as the law of degradation of energy and matter. Max Planck was the first to bring to the attention of physicists that this statement does not convey the complete meaning of the law. "The real meaning of the second law has frequently been looked for in a 'dissipation of energy.' This view, proceeding, as it does, from the irreversible phenomena of conduction and radiation of heat, presents only one side of the question," wrote the father of quantum mechanics.[3]

The side that has been emphasized and publicized in the literature is the one that associates the Second Law with the availability of energy and natural resources. Because the First Law states that energy is constant, and because the Second Law states that energy dissipates in energy transformations, the conclusion has been drawn that we are running out of energy and resources, as they are being transformed into less available, more degraded forms.

Many books and reports espouse this worldview, although they may not necessarily mention the Laws of Thermodynamics directly. The Club of Rome's *The Limits to Growth* and the Carter administration's *The Global 2000 Report to the President* are examples of publications that portray a world whose finite resources are being degraded and depleted at a rapid pace, and whose population is ever increasing, suggesting that if present trends are not reversed, then major disturbances await us—even to the point of the collapse of the world economic system and civilization as we know it.[4]

The conclusions reached in *The Limits to Growth* and *The Global 2000 Report* have been attacked on technical grounds. Opponents say that the data used in their models are unreliable, that there is no convincing information concerning Earth's reserves of fossil fuels such as petroleum, coal, or natural gas, and that the uncertainties are even greater when it comes to prospects for the so-called renewable forms of energy, directly or indirectly derived from the Sun.[5] Economist Julian L. Simon wrote, "technological forecasts of resource exhaustion are often unsound and misleading," because the total quantity of natural resources is unknown. Even if the quantity were known, it would not be economically meaningful since we have the capacity to develop new ways to meet our demands—for example, by exploiting "low grades of copper ore previously thought not usable, and by developing cheap atomic power to help produce copper." He concluded that the "existing 'inventory' of natural resources is *operationally* misleading," because physical measurements cannot define what humankind will use in the future.[6]

The fact that "cheap atomic power" has not come about, nor that there is any indication that it ever will, is another issue. The principle that Simon used to reject the reliability of reports like *The Limits to Growth* is an old one. It is called the principle of substitutability of materials. If humanity runs out of material X, material Y will be used to replace it. In 1976, scientists H. E. Goeller and A. M. Weinberg formalized this concept: "We now state the principle of 'infinite' substitutability: With three notable exceptions—phosphorus, a few trace elements for agriculture, and energy-producing fossil fuels (CH_x)—society can subsist on inexhaustible, or near-inexhaustible minerals with relatively little loss of living standard. Society would then be based largely on glass, plastic, wood, cement, iron, aluminum, and magnesium."[7]

Opponents of *The Limits to Growth* report bring up yet another point. They remark that the computer model "tacitly assumes that the Earth is a closed system," and that the only energy and material resources available to us will be those of planet Earth.[8] "Then there is outer space and the planets," wrote Simon. In support of his point of view, he related what some scientists had envisioned at a meeting of the American Association for the Advancement of Science. Peering into their crystal balls, "the scientists concluded that space colonization is inevitable—and sooner than we think. . . . The most important leap into space will begin in 1980, when the space shuttle takes its first payloads into Earth orbit. . . . Mining the moon can begin in 1990. The material from 50 million tons of moon rocks can be used to make solar-power satellites that will provide all of the Earth's energy needs by 2000."[9]

The space shuttle became operational a few years behind schedule, and has walloped most of NASA's budget. It has not yet begun to solve the problem of the availability of energy and resources because it is struggling to solve its own internal problems, while consuming a lot of resources all by itself. Only time will tell whether 50 million tons of moon rocks will ever meet all the energy needs of the world.

A further rebuttal to the *Global 2000 Report* came from a book composed of a compendium of studies by academics and scientists covering almost the full spectrum—from population trends and agricultural prospects to energy, minerals, and the environment. The editors of the book, *The Resourceful Earth: A Response to Global 2000*, state categorically that the *Global 2000 Report* is "dead wrong."[10]

One of the major disagreements between these two reports revolves around the availability of energy. *The Resourceful Earth* states, "The prospect of running out of energy is purely a bogeyman. The availability of energy has been increasing."[11]

The Second Law makes a specific statement about the availability of energy and matter. Four decades ago, physicist R. B. Lindsay put it this way: "every naturally occurring transformation of energy is accompanied, somewhere, by a loss in the *availability* of energy for the future performance of work. . . . If the essence of the first principle in everyday life is that we cannot get something for nothing, the second principle emphasizes that every time we do get something we reduce by a measurable amount the opportunity to get that something in the future, until ultimately the time will come when there will be no more 'getting.'"[12]

This description of the way the world works, although correct, tells only one side of the story—the side that emphasizes the importance of the availability of energy. While it is true that the availability of energy is decreasing in the universe, and also in our own closed thermodynamic system—Sun and Earth—this concept should be used with care, as it may not be meaningful in everyday life. For example, would we want to have at our disposal all the available energy of the big bang—all 10^{32} K condensed in a space of 10^{-33} centimeters? Could we extract any useful work from this "inexhaustible" source of energy? "One absurdity usually not discussed even among professionals," wrote Edward Teller, "is that in spite of temperatures greatly exceeding those of Dante's inferno, and in spite of the enormous density of energy present, none of the energy would have been available for the purpose of performing useful functions."[13]

While the Second Law specifies that the potential for performing work diminishes with the passage of time, it tells us nothing about the type of work that can be accomplished with a given kind or a given amount of energy. For instance, at the big bang there was enough energy available to create the Cosmos. The available energy, although enormous, was not in a form that could maintain life as we know it. It took a few billion years and considerable dissipation of energy before life could appear and evolve on Earth. Although the available energy of the universe is still inexhaustible for all practical purposes, most of it is in a form that cannot be used to perform useful work. The majority of it is in the form of kinetic energy in the distant galaxies, which are receding at extremely high velocities. This energy, as well as most of the universe's energy, is simply unavailable to us.

Humankind has plenty of available energy from coal, wood, oil, natural gas, nuclear energy, and—most of all—the Sun's radiation. For us, the most important source of energy is and will be our Sun. It is ample and, for all practical purposes, inexhaustible. As Teller reminds us, "The sun has been shining for five billion years and there is no reason why it should not continue for five billion more."[14]

Entropy: The Supreme Manager of All Natural Processes

Suppose we had all the energy and natural resources we wanted to have. Would the Second Law of Thermodynamics be overtaken? Partially. One side of the law would be taken care of—available energy and resources would no longer be a problem for us. The other side would still be there, however, the side that exacts a penalty in all energy-matter transformations through increases in the entropy of the thermodynamic system.

Take, for instance, the London pea-soup fogs of the 1950s, caused by the burning of abundant coal. One of these fogs killed thousands of Londoners. They did not die because of a lack of available energy. They choked from sulfur dioxide generated by the burning of fossil fuel. "Paradoxically, I see our greatest dangers for centuries to come not in shortages of energy but in excessive use of it, either for our own individual lives or for the manipulation of the environment," remarked ecologist René Dubos in 1981.[15] Every time we perform some work or transform energy-matter from one kind to another, we pay our dues to the Second Law.

Max Planck is famous for his theoretical investigations on radiation, which gave birth to quantum mechanics and a new era of modern physics. His doctoral dissertation, however, was on the Second Law. He always considered thermodynamics his "own home territory," where he felt "on safer ground."[16]

Planck remarked that the Second Law has often been stated in this fashion: mechanical work can be completely transformed into heat, while the opposite process must be incomplete. The incomplete transformability of heat into work came from Carnot's principle, which states that in heat engines, some quantity of heat must always be dissipated into the lower-temperature reservoir (the environment). Thus the mechanical work output can never be equal to the heat input.

"This is quite correct," wrote Planck, in certain cases (in heat engines), "but it by no means expresses the essential feature of the process." To prove his point, he gave a simple example: the expansion of a perfect gas at constant temperature. If we take a perfect gas and connect it to a heat-reservoir of higher temperature, he said, and then allow the gas to expand isothermally, doing external work, "the temperature of the gas, and at the same time its internal energy, remains unchanged, and it may be said that the amount of heat given out by the reservoir is completely changed into work without an exchange of energy taking place anywhere. Not the least objection can be made to this. The proposition of the 'incomplete transformability of heat into work' cannot be applied to this case."[17]

Planck's example is an uncommon one in everyday experience. In our activities, we do not transform heat into work through the expansion of perfect gases at constant temperature. Nonetheless, Planck's example shows clearly that degradation of energy cannot be the essence of the Second Law. As noted physicist Arnold Sommerfeld pointed out, "In this process energy is not degraded but quite to the contrary, it is ennobled (heat completely transformed into work)."[18]

Sommerfeld agreed wholeheartedly with Planck that the gist of the Second Law lies in the statement that in all natural processes the quantity called entropy always increases. In support of Planck's point of view, he wrote: "Planck opposes (and rightly so) the view of certain physicists that the essence of the Second Law consists in the statement that energy tends to degrade. Evidently an increase in entropy causes in many cases a decrease in the available temperature difference and hence also in the availability of work. Planck quotes the obvious example in which heat is transformed into work completely. . . . In our and in Planck's opinion, the essence of the Second Law consists in the existence of entropy and in the impossibility of its decreasing under well defined conditions."[19]

Sommerfeld was no ordinary physicist. During the first third of twentieth century, "he was arguably the best theoretical physics teacher. His textbooks are among the finest to this day," notes the biographer of Niels Bohr.[20] Although he did not receive the Nobel prize, four of his doctoral students did—Hans Bethe, Wolfgang Pauli, Peter Debye (in chemistry), and Werner Heisenberg.[21]

The idea of dissipation of energy was initiated and popularized by Lord Kelvin in his landmark paper, "On a Universal Tendency in Nature to the Dissipation of Mechanical Energy."[22] In it, he gave many examples of physical processes where "*dissipation* of mechanical energy" occurs, making it impossible to obtain "perfect *restoration*" of the initial conditions, meaning that physical processes are irreversible—the central proposition of the Second Law of Thermodynamics.

Kelvin came up with the idea of dissipation of mechanical energy from Carnot's work on heat engines.[23] And because Carnot's work was instrumental in the formulation of the Second Law, Carnot's maximum efficiency principle has often been regarded—incorrectly—as being the Second Law itself.

Neither Planck nor Sommerfeld was arguing against Carnot's maximum efficiency principle, which is well established when applied to heat engines. They were pointing out, however, that the universe (or Nature) is not a heat engine exclusively but a thermodynamic system in which many diverse thermodynamic processes take place. For example, many processes occur in Nature isothermally—at constant temperature—in which case Carnot's principle is not applicable but the Law of Increasing Entropy still is. If we look at the nuclear processes occurring on Earth, in the Sun, and elsewhere, we find no connection with Carnot's principle because nuclear reactions are not caused by temperature differences between various nuclei. Clearly, Carnot's principle cannot be the essence of the Second Law—a law that governs *all* natural processes.

Numerous physical, chemical, and biological processes occur—on Earth and in the universe—where Carnot's efficiency principle is not a factor, while the Law of Increasing Entropy is. Take the case of "fuel cells," an idea more than a century old. In fuel cells, when streams of hydrogen gas and oxygen gas are made to combine to form water, their chemical energy is transformed directly into electrical energy, without an intermediate thermal stage. A fuel cell is analogous to a battery in which chemical energy is directly converted into electrical energy. "The efficiency of fuel cells," writes Teller, "is not limited in an obvious way by Carnot's principle or its formulation in the second law of thermodynamics because no temperature changes need be produced."[24]

A fundamental question remains: What happens to the Second Law when chemical energy is transformed into electrical energy, as in batteries and fuel cells? Does it mean we have neutralized the Second Law? Not at all. As Planck and Sommerfeld pointed out, the essence of the Second Law does not lie in Carnot's efficiency principle. It lies only in one universal principle, *the inexorable increase of entropy*, which is locked within every process.

Because the entropy side of the Second Law has often been ignored or given less attention, the so-called renewable energy sources derived from the Sun have been hailed as benign alternatives to oil, coal, and nuclear plants. However, even the "ecologically favored" solar sources of energy have serious drawbacks. In a report issued by the National Audubon Society, physicist Larry Medsker, after surveying nine renewable energy sources, found that "all have potentially unwelcome, occasionally even hazardous, side effects." For example, the burning of wood on a large scale not only can deplete forests but also increases air pollution. As Russell W. Peterson—past president of the Audubon Society—summarized, "Even with solar energy, there's no such thing as a free lunch."[25]

Not only is there no free lunch, but entropy is produced in all energy usage. "Any use of energy has an environmental impact of some kind," writes physicist Robert H. Romer in *Energy: An Introduction to Physics*, adding, "each particular kind of energy use has its own special effect on the environment in addition to the generation of heat, not only on scenery and wildlife but also on the health of the human inhabitants. Some of these effects are well known; there are undoubtedly others of which we are still ignorant."[26]

The statement that "any use of energy has an environmental impact of some kind" applies also to solar energy—we still have to account for the entropy generated by the processes that transform electromagnetic energy into other forms of energy. Consequently, we must not focus our attention solely on the "availability of energy" side of the equation, as many technologists, economists, and even some environmentalists do.

While Carnot's efficiency formula and availability of energy remain useful concepts, especially in engineering fields, the concept of entropy maintains a higher position because of its universality. Unquestionably, the Law of Entropy is the Supreme Manager of all natural processes.

It is useful here to quote Robert Emden, "whose deep understanding of thermodynamics," in the words of Sommerfeld, "has withstood the test of time."[27] In 1938, writing in *Nature*, Emden expressed his view on the relative rank of energy and entropy:

As a student, I read with advantage a small book by F. Wald entitled "The Mistress of the World and her Shadow." These meant energy and entropy. In the course of advancing knowledge the two seem to me to have exchanged places. In the huge manufactory of natural processes, the principle of entropy occupies the position of manager, for it dictates the manner and method of the whole business, whilst the principle of energy merely does the bookkeeping, balancing credits and debits.[28]

The Greenhouse Effect

Nature has provided us with an ample supply of readily available energy in the form of coal, natural gas, oil, and gasoline. Since the industrial revolution some two centuries ago, there has been a dramatic increase in the burning of these fossil fuels. All fossil fuels are either carbon—coal and graphite—or carbon-hydrogen compounds, like methane, ethane, propane, butane, and octane. When they undergo complete combustion with oxygen, carbon dioxide (CO_2) and water enter the atmosphere. Carbon dioxide, an innocuous gas, remains in the environment for decades. Eventually it is dissolved in the ocean and precipitated on the bottom as limestone.

Due to the accelerated burning of fossil fuels, mainly oil and coal, the amount of carbon dioxide spewed into the atmosphere is increasing steadily. Because humankind is systematically destroying rain forests, which absorb CO_2 naturally, the concentrations of CO_2 in the atmosphere have increased by 30 percent since the industrial era. In 1998, CO_2 concentrations reached the highest level in 420,000 years.[29] But why worry about it if this chemical is not poisonous? In this case, the problem lies not in the chemistry of things, but in the physics. Carbon dioxide is a gas that freely admits sunlight into the atmosphere, but prevents lower-energy infrared heat radiation from escaping back into space, thus disrupting the planet's temperature regulation system. This phenomenon is called the "greenhouse effect." CO_2 acts as a thermal blanket by trapping the heat radiation into the atmosphere, causing Earth's temperature to rise.

There are other gases that contribute to the greenhouse effect. These include methane, nitrous oxides, and chlorofluorocarbons (CFCs). The more wood we burn, and the more cattle we raise, the more methane is produced. Chemical fertilizers and car emissions generate nitrous oxides. The synthetic CFCs are found in refrigerators and aerosol propellants.

A warmer climate would have many side effects. For example, higher temperatures make water expand. Thermal expansion has already raised sea levels by four inches since the turn of the twentieth century. The equatorial regions can absorb substantial amounts of heat by evaporating water, but the polar regions do not have such a neat way to dispose of great quantities of additional heat energy. The poles would warm up and the polar ice caps would partially melt, inundating coastal lowlands. Nations at higher latitudes would experience greater heating than equatorial regions. Continental interiors would suffer more from heat than coastal regions.

"That man affects his environment, and that we should take these environmental changes into account, is generally accepted today," wrote Edward Teller in 1979. "It is by no means clear which will happen first: the exhaustion of our fossil fuel reserves, or an overproduction of carbon dioxide due to the accelerated burning of these fuels."[30] More than two decades have gone by. The debate over the exhaustion of available energy has subsided, while the greenhouse effect is now attracting much attention.

Meteorology is a tricky business. There are many other effects that could counteract the warming effects of the carbon dioxide blanket. However, the physics behind CO_2's greenhouse effect is irrefutable: the more CO_2 in the atmosphere, the more heat will be trapped. Today, only a handful of scientists dispute the fact that the amount of CO_2 in Earth's atmosphere has increased and is still rising, and that "the human race is thus conducting a dangerous experiment on an unprecedented scale."[31]

More and more scientists are beginning to believe that the greenhouse effect has begun. In 1989, James Hansen of NASA's Goddard Institute for Space Studies declared the evidence is "pretty strong that the greenhouse effect is here."[32] The 14 hottest years on record have occurred since 1980, and 1998 was the hottest ever recorded.[33] Mountain glaciers are melting; coral reefs are dying; coastlines are eroding; ocean levels are steadily rising and inundating low-lying areas of islands; homes, roads, bridges, and plantations are washing away; and unusually severe weather has become more frequent. Moreover, insects that carry tropical diseases have begun to move into regions that were once too cold for them.[34]

Climate changes have, of course, occurred throughout time due to thermodynamic changes; the current concern is their rate of change. We are contemplating much higher rates of climate change than at any time in human history.[35] Consequently, ecosystems will not be able to adjust quickly enough, and the faster things change, the higher the probability that the overall impact will be disruptive.

Thermodynamics tells us that heat is a form of energy. Thus, higher temperatures mean there is more available energy driving our planet's climate system, which in turn means more evaporation, destructive storms, and flooding. While some individuals, organizations, and industries are denying the existence of global warming, the insurance industry is not; it has been burdened with billions of dollars in unprecedented claims from weather-related damage.[36] In 1998, the insurance industry worldwide registered $92.9 billion in weather-related damages, three times the figure in 1997 and half again as much as the previous record of $61.7 billion in 1996. Indeed, the figure for 1998 alone exceeded the $78.4 billion in losses for the entire 1980s.[37]

The disasters in 1998 killed about 32,000 people, and drove another 300 million from their homes. Hurricane Mitch and the flooding of China's Yangtze River were among the worst calamities. Mitch, the deadliest Atlantic storm in two centuries, took about 11,000 lives when it barreled into Honduras, Nicaragua, Guatemala, and El Salvador, and left a million people homeless. It also wiped out about 100 bridges and half of all food crops. Flooding of the Yangtze River, the costliest disaster of 1998, will cost China $30 billion. The flooding dislocated more than a fifth of a billion people, inundated 25 million hectares of cropland, and caused 37,000 deaths.[38]

The 1999 Atlantic hurricane season witnessed five Category 4 storms—the first time on record. Category 4 is the second most powerful storm level, with winds of 131–155 miles per hour capable of destroying mobile homes. The 1999 hurricane season was a continuation of a period of intense hurricane activity that began in 1995, making it the busiest five-year period for hurricanes in history.[39]

As more greenhouse gases are emitted, and more forests de-
nuded, the planet's thermodynamic clock keeps ticking away. Even
if CO_2 emissions stopped completely today, some warming is un-
avoidable because of the greenhouse gases already in the atmosphere. CO_2 tends to stay in the atmosphere for about a century.[40] It
is like a genie let out of the bottle. We have to remind ourselves that
we cannot come back to today's environment if we do not like the
new one: the Second Law locks the door behind us.

High Tech's Environmental Entropy

*In the face of the fact, the scientist has a humility almost
religious.*[41]

—Percy Bridgman, *Nobel laureate in physics*

Because the United States is becoming more and more of an information society, as distinct from an industrial society, some people
are led to believe that environmental disorders will be decreasing:
pollution-rich smokestack industries will shut down in favor of
nonpolluting computer, microchip, and communication industries.
Of course, the global entropy production would not be declining,
since these smokestack industries will have moved to the developing countries. However, the problem remains that computers generate their own environmental entropy.

How does silicon, copper, or any other raw material become a
microchip? Through complex thermodynamic processes. If we follow all the processes from beginning to end, we discover that many
chemicals are used—some of them toxic—in the making of microchips. The more of them we produce, the more we will need the
gases, acids, dopants, and solvents used in their manufacture. What
happens to these chemicals after their use?

In 1984, the inhabitants of Silicon Valley—the bastion of microelectronics technology—were shocked to discover that microchips

create environmental disorders, though not visibly like sulfur oxides in the air. "Silicon Valley's electronics factories, once thought to be almost antiseptically safe," wrote the *Wall Street Journal*, "turn out to have a messy underside: toxic chemicals used in some high-tech manufacturing processes are seeping out of storage tanks and into local water supplies. For a business usually considered clean and free of health hazards—especially in comparison with heavy industry—the recent revelations are both surprising and disillusioning."[42] Thermodynamically speaking, it would have been a miracle, a violation of the Second Law of Thermodynamics, if it were otherwise—that is, if humans could convert silicon to microchips without increasing the entropy of the environment.

In just two years, some 70 cases of toxic leaks became known in the Silicon Valley. In 1984, dangerous amounts of a computer-chip degreasing agent were discovered in nonpublic tap-water wells.[43] And in 1986, traces of chemicals leaked by semiconductor manufacturers were found for the first time in Silicon Valley's deepest aquifers, from which public water supplies are taken.[44] There are now "178 sites in Santa Clara County where high-tech firms have polluted the local groundwater," reports Aaron Sachs of Worldwatch Institute.[45]

The shocking turn in Silicon Valley's image began in 1981 when residents of a San Jose neighborhood near a semiconductor plant started noticing a high level of birth defects. The following year a high-tech firm revealed that an underground storage tank was leaking. In 1985, the California Health Department released a two-year study showing that "the number of birth defects in that area was more than triple that of a control area."[46]

The microelectronics industry spews huge quantities of "reactive organic gases" into the air daily. These compounds become involved in the formation of photochemical smog. "The 'clean' image of the microelectronics industry is misleading," wrote Dr. Joseph LaDou in *Technology Review*, adding: "It is not an overstatement to say that the semiconductor industry provides a complete spectrum of the occupational hazards found in most industries," including exposures to chemicals, gases, and metals.[47] In so-called clean rooms, where intricate circuitry is etched onto the silicon wafers, "production workers never breathe in any dust, but they are regularly exposed to known carcinogens like dichloroethylene."[48]

Computers and other high-tech equipment are generating yet another environmental entropy, known as "hash" or "noise." Electrical devices, whether they be microwave ovens, computers, or cathode ray tubes, give off impulses that travel through the air as radio waves. This invisible electromagnetic pollution—an unwanted by-product of the electronic age—is blamed for disrupting operations from jet flights to drunk-driving tests. For example, during the first landing of the space shuttle *Columbia*, stray radiation from the cameras of TV crews covering the landing interrupted communications between the astronauts and ground controllers. The electronic hash, also called electronic smog, is increasing steadily worldwide.[49]

The radiation emitted by high-tech devices including video-display terminals has an effect on our health.[50] Eyestrain is a common complaint of those who spend many hours staring into the computer screens.[51] In *Cross Currents*, Dr. Robert O. Becker speaks out on the health hazards of a wide spectrum of modern equipment—from television sets to fluorescent lights to microwave ovens to hair dryers—that emit electromagnetic radiation, which he calls "electropollution." He points out that the phenomenon of life is controlled by the same forces of Nature that have shaped the universe, and that life from its inception has been dependent on Earth's natural electromagnetic environment. Today, this natural environment is submerged beneath a torrent of electromagnetic fields that never existed before. The human-made electromagnetic environment is composed of radio and TV signals, microwave transmissions, high-power tension lines, radar, and other electromagnetic currents and waves that crisscross our world. Becker writes, "many studies have shown that this [electromagnetic] radiation, which was previously thought to be innocuous, may be extremely hazardous to our health. In fact, it has been linked to the increased incidence of certain cancers, birth defects, learning disabilities, and mood changes."[52]

Can We "Control" Natural Processes?

What can be controlled is never completely real; what is real can never be completely controlled.[53]
—Vladimir Nabokov

The idea of "control" dates back to antiquity. In time, the concept has evolved both in meaning and in scope. When machines started to play a large role in our lives, interest in the subject of control took on a new impetus. For instance, when we began to build steam engines and electrical motors, we realized they needed some kind of control mechanism because the loads change considerably during operation due to environmental fluctuations. A machine or an engine needs a feedback loop or a control mechanism to keep it running smoothly despite variations in the load.

Such a controlling device was invented for steam engines by James Watt himself, who quite appropriately called it a governor.[54] The device, which was attached to the engine, used a small amount of the energy being transformed by the engine to control the rate of the energy transformation. The governor ascertained when the load demanded more or less power, and communicated the relevant information to the throttle, instructing it to pass more or less steam as needed. This process is known as feedback control.

More recently, the field of automatic or feedback control came to prominence with the publication of Norbert Wiener's book, *Cybernetics, or Control and Communication in the Animal and the Machine* (1948).[55] There, the MIT mathematician brought together, for the first time, references to a host of phenomena from different fields of science and technology where the concept of control, particularly involving feedback, was central. Since then the subject has proliferated. It has entered almost every aspect of human activity—from psychology and biology to engineering and philosophy.

Wiener was not the first to use the term cybernetics, however. That distinction belongs to the French physicist André-Marie

Ampère. In 1834, he introduced the word cybernetique to denote the art of government and control in general.[56] Cybernetics comes from the Greek "kubernetes," meaning a pilot or navigator, the person who controls the direction of a ship. This was the original meaning of the word. The ancient Greeks called "Cybernesies" the festivals that were instituted by Theseus in memory of the pilot who had guided his vessels in his Cretan expedition.

Soon the original meaning was elaborated and took on extended significance, even among the Greeks. The *Grande Larousse Encyclopedique* notes that "the Greeks generalized the art of control and the word 'cybernetics' is even found in several dialogues of Plato. This celebrated philosopher indeed amplified the meaning of the verb 'to govern' and provided many different examples, all the way from the steering of a ship to the driving of a chariot and finally to the governing of people. On his side Xenophon sought to study in systematic fashion the art of government in the political sense."[57]

Throughout history, the idea of control has attracted the attention of people from a broad spectrum of interests, from philosophy and government to science and technology. It is not surprising that the idea has had great appeal. We all want to be in control of the situations, events, and environments around us. If we were in government, we would like to be able to control people's behavior; if we were in economics, we would want to have the capability to control the economy and steer it in the "appropriate" direction; if in biomedicine, we would like to control such things as diseases and the aging process. The list is long, because we all have a certain desire to control situations that affect us, to be in control of our lives and our destiny. As civilizations have advanced socially, scientifically, and technologically, this list has not only grown proportionately but has also become increasingly ambitious in content. We have reached a point where the list contains not only such things as machines, humans, and economies, but also the forces of Nature and the natural trend of evolutionary processes.

There is no question that science and technology have given us the means to "control" a wide range of processes, including some of Nature's own. Consider Nature's heat and its flow. We know that heat flows naturally (by itself) from a hotter to a colder body. Can we reverse this natural direction of heat flow? Yes, by running a heat engine in reverse, as in refrigerators and air conditioners. A heat engine takes in heat from a high temperature source, converts part of it to other forms of energy—such as electricity—and ejects the difference as heat in the exhaust at a lower temperature.

A refrigerator, on the other hand, takes energy input in the form of mechanical work (W) or electricity, uses it to extract heat (Q_1) from an object at a low temperature (a bottle of beer or champagne), thereby making it cooler, and ejects the sum ($Q_2 = Q_1 + W$) as heat at a *higher* temperature. Most refrigerators have coils of tubing in the back that are warm to the touch. This is where the heat produced gets delivered to the room, which must absorb both the heat "pumped out" of the refrigerator and the heat equivalent of the work done by the motor. In this example, we did control Nature's heat flow, but we ended up with more heat to contend with. Now if we want to get rid of this extra heat, we can install a window air conditioner, which must pump out this heat, the heat in the room, and the heat equivalent of the work done by its own motor to the air outdoors. If we went outside, we would not only feel the heat coming out of the air conditioner, but would also hear the motor's distinct noise.

We lowered the heat (entropy) inside the room and the refrigerator, but we did so at the expense of increasing the overall heat—entropy—and the level of noise. We still have the big "W" (the electricity) to account for: that is, we have to deal with the additional heat and entropy generated at the place where the electricity was produced. If nuclear fission is the local source of energy, we have to deal not only with the heat produced by the nuclear plant but also with the highly radioactive by-products of the nuclear reactions. Because we cannot "control" the output of these radioactive by-products, since they are direct participants in the nuclear reactions, the most we can do is to "monitor" their radioactivity through elaborate monitoring equipment. And should the processes that monitor and control the nuclear chain reactions malfunction for some reason or another, causing radiation to escape and endangering people, livestock, agriculture, and water supplies, we have on our hands a lot more monitoring and controlling to do. These activities require expertise in such diverse disciplines as chemistry, biology, and immunology.

Of course, not all power plants use nuclear fission to generate electricity. If fossil fuels were used, then we have other emissions to contend with, such as gases that contribute to acid rain and the greenhouse effect. Moreover, as noted earlier, the global warming phenomenon affects weather patterns through heat waves, droughts, floods, and hurricanes.

We have now reached a point where we need to extend our controlling mechanisms into all corners of the world, as the problem has grown to the level where it requires a global response. To control the behavior of humankind, *à la cybernetique*—that is, through information feedback and control—is no trivial task, since it involves the control of at least 6 billion people with different ideologies, religions, and cultures, all with states of minds and economic situations in constant flux.

The Laws of Thermodynamics tell us that the more things we try to control, the more entropy we generate, requiring the institution of even more controlling mechanisms. In our refrigeration example, all we wanted to do was to control Nature's heat (energy) flow for our comfort and well-being. We did not intend to create worldwide disturbances, but we did; we increased not only the overall heat but also such things as radioactivity, carbon dioxide, and acidity in rain, which in turn demand from us even more monitoring and controlling activities.

In his 1989 book *The Control of Nature*, writer John McPhee, a winner of the Pulitzer prize, documents a few situations where modern humans have been "engaged in all-out battles" with Nature. He was motivated by a sign he happened to see at the University of Wyoming's engineering building, where words etched in limestone read: "STRIVE ON—THE CONTROL OF NATURE IS WON, NOT GIVEN."[58] The control of the Mississippi River is one such project that attracted McPhee's attention. After more than 100 years of official U.S. government assurances that Nature's rivers can be controlled with military precision by the Army Corps of Engineers, the Mississippi River was still not behaving the way the corps wanted. The corps, which has made a habit of declaring final victory over the forces of Nature every few years, was being overwhelmed by the Mississippi's conduct. As *U.S. News & World Report* points out, the Mississippi levees to control the floods "stood 6 feet high in 1850; a century later they had risen to 30 feet, and the floods continued, each time more devastating, as the river towered ever higher above the land below."[59]

Whether technologies are used to subdue Nature or to control the behavior of machines, the Laws of Thermodynamics are equally applicable. All activities, including the controlling ones, require some energy-matter transformations, bringing them under the surveillance of the Laws of Thermodynamics. As physicist R. B. Lindsay remarks in *The Control of Energy*, "the ultimate control of energy or everything else in human experience, for that matter, is exercised by Nature itself and described by the principles of thermodynamics."[60]

It is true that cybernetic devices—like thermostats—do not need much energy to operate. However, they do require energy and matter to produce, and they do break down and malfunction. Consider, for example, the microchips embedded in modern appliances, gadgets, automobiles, and other machines to monitor and control their behavior. Although these cybernetic devices are very small and becoming even smaller, their production calls for, among other things, chemicals that require monitoring. If these chemicals leak into our water supplies, we must then set up elaborate quality control mechanisms, as our very livelihood depends on the availability of clean water. Many of today's cybernetic products run on batteries, which not only require energy-matter to produce, but also hold chemicals that must be properly disposed of and monitored.

We have also made great efforts to control the biological environment around us. We have introduced chemical after chemical and superweapon after superweapon, in an attempt to control the biological world of insects and pests. However, after half a century of intensive work the insects are still thriving, requiring the introduction of even more chemicals to control their biological processes, while all along the pesticides have increased the disorder of our thermodynamic environment. This in turn has led to the introduction of yet another controlling mechanism, a bureaucratic one—the Environmental Protection Agency, whose function is to control the usage of chemicals deemed dangerous to our health and to provide the funds to clean up the environmental disorders generated by chemicals.

In the area of medicine, we have also introduced drug after drug and magic bullet after magic bullet in an attempt to control a broad range of biomedical disorders, from infectious diseases to the aging process itself. Immortalists believe that through biomedical interventions we will eventually eliminate killer diseases and conquer death. They point out, for example, that the discovery of antibiotics and the development of vaccines have led to the conquest of various killer diseases. Smallpox was officially declared eliminated world-wide in 1979. Polio has all but vanished in the United States, and measles cases have dropped considerably. However, after more than a half century of massive attack on microbes, bacteria—having acquired considerable immunity to our wonder drugs—are flourishing and are attacking us with greater impunity.

When antibiotics were first developed, death from sepsis—systemic blood poisoning that stems from bacteria infections—dropped dramatically as the use of penicillin, streptomycin, and the sulfanilamides became widespread. But a curious phenomenon unfolded. As time when on, sepsis patients began to need an increasing dosage of penicillin to be saved from the deadly disorder. Sepsis victims, who might have been saved by a few hundred thousand units of penicillin earlier, needed tens of millions of units some 20 years later. By 1967, the death rate from sepsis had jumped back to pre-antibiotic levels. The explanation for this phenomenon is straightforward: antibiotic resistance and adaptation.

As early as 1942, Sir Alexander Fleming, who had discovered the antiseptic properties of penicillium mold some 14 years earlier, warned of the emergence of antibiotic resistance among the staphylococci, the bacteria that cause boils and other skin disorders. "But because of the dramatic early effectiveness of each new generation of antibiotics, drug companies continued their quest for more," writes experimental pathologist Marc Lappé. "Each new antibiotic was heralded as a wonder drug, and an end was predicted to the infectious diseases that had plagued previous generations."[61]

Today, many of the common pneumonia-causing bacteria are no longer responsive to penicillin. The ancestors of these infection-causing microorganisms gradually adapted through the powers of mutation and became resistant to antibiotics. Moreover, they have transferred genetic knowledge from one group to another. Yesterday's bugs have evolved to become today's "superbugs."[62] And today's superbugs will become tomorrow's super-superbugs. Thus, the need for new, more potent antibiotics becomes a never-ending battle. We must constantly try to outsmart the clever and adaptive germs.

The problem of drug resistance is not confined to a few germs but "spans an entire spectrum of disease-causing microbes," which include those responsible for tuberculosis, gonorrhea, and meningitis.[63] "Twenty years ago, many believed infectious diseases would be conquered," recalls Dr. John Bartlett, chief of the infectious diseases unit at Johns Hopkins Medical School. "But not only are they not gone today, these diseases also seem to have thrived."[64] As Lappé remarks, "the more vigorously we have assailed the world of microorganisms, the more varied the repertoire of bacterial and viral strains thrown up against us."[65]

The principal reason why antibiotics, or drugs in general, have not controlled or eliminated infectious diseases is the same reason why pesticides have not eradicated undesirable insects. Every time we have introduced a new weapon against insects, we have become elated by the early successes of the weaponry and have declared self-confidently, but prematurely, that victory is at hand. For example, at one time malaria was thought to be beaten by the "super-weapon" called DDT. The preliminary data looked promising, and the disease appeared exterminated.

But throughout the world, malaria has bounced back with new strength, although it seemed almost conquered in the decade or so after World War II. Since then, mosquitoes have become resistant to pesticides, and the malaria parasite has learned to cope with some of the more widely used drugs. In some areas it is even more dangerous and widespread than ever before. (One estimate puts the number of mosquitoes worldwide at around 100 trillion.)[66] Malaria now kills approximately twice as many people worldwide as AIDS, and sickens as many as half a billion.[67] Mosquitoes in diverse regions of the United States carry the disease as well as the dangerous encephalitis virus.[68]

With all our interventions to control the aging process, we have been unable to do so; all of us are still aging—increasing our entropy—irreversibly. Because every drug brings with it some side effects, we have introduced yet another controlling mechanism, a bureaucracy called the Food and Drug Administration (FDA), to control the spread of ineffective and dangerous drugs. Even with this controlling entity in place, many FDA-approved drugs have proved to have severe side effects, some deadly. "Since 1980, 13 drugs have been recalled because they proved to be unsafe. Many others remain on the market despite potentially fatal side effects," pointed out *U.S. News & World Report* in 1997, adding: "The General Accounting Office estimates that 51 percent of FDA-approved drugs have major adverse effects that aren't detected until the drugs are taken by the general public. As many as 140,000 people a year die of reactions to prescription drugs."[69]

The situation is no different when it comes to controlling and governing people. "The relations of people with other people," wrote André-Marie Ampère in 1834, "are only the least part of the objects to which a good government should devote its attention. The maintenance of public order, the execution of the laws, the just apportionment of taxes, the choice of men for employment, and everything that can contribute to the amelioration of the state of society demand its never ending concern."[70] However, all these cybernetic activities require energy. From where does the government get it? From the same place that Watt's governor did, the very engine that the cybernetic device was designed to control: the people.

While Watt's governor took a minute amount of the engine's energy to control its behavior, social governments take a big chunk of people's energy to perform cybernetic tasks. Like other controlling mechanisms, they occasionally falter, necessitating additional regulatory mechanisms to control the behavior of the governments themselves.

There is a fundamental difference between Watt's cybernetic device and social governments. The steam engine that Watt's governor controlled was designed by Watt himself, who knew not only the physical characteristics of the engine, but also how it would behave under various conditions. This is true for most if not all control mechanisms. Governments, however, do not have a precise model of how millions of people behave under various situations, nor do they have the capability to predict tomorrow's situation accurately. Not surprisingly, governmental steering mechanisms often do not work as advertised. In addition, people are aware that government actions affect them. Consequently, they monitor government activities, setting up lobbyists and other special groups to steer governments in their own direction.

A question of natural philosophy remains: How is the universe governed, and its behavior controlled? Does the universe have an elaborate mission control room that controls the movement and behavior of the galaxies, stars, planets, meteorites, and everything in between? Not to our knowledge. The overall behavior of the universe is controlled by the Forces and the Laws of the Universe, which on Earth we call the Laws of Nature.

In our solar system, the planets' trajectories and movements are governed by the gravitational forces and the law of gravity. The nuclear reactions in the Sun come about through the nuclear forces in the nuclei, and the resulting thermal electromagnetic radiation obeys the laws of electricity and magnetism.

The individual Laws of Nature control the behavior of a particular set of natural phenomena. As physicist Henry Margenau has pointed out in *The Nature of Physical Reality*, however, there are other laws or "principles operating above the plane of laws, principles unable in themselves to generate the laws but able to give them scope and substance. They are often called the Laws of Thermodynamics for historical reasons." Because "they function as super laws," they are capitalized.[71] By working above the plane of laws, they act as the Supreme Governor of all events and processes. They are the true cybernetic Laws of the Universe.

Chapter 8

Economics, the Environment, and the Laws of Thermodynamics

From the viewpoint of thermodynamics, matter-energy enters the economic process in a state of low *entropy and comes out of it in a state of* high entropy.[1]
— Nicholas Georgescu-Roegen

Economic Theories

Economists, with rare exceptions, have formulated their theories and expounded their sociological philosophies without much regard to the Laws of Nature. And whenever they have linked their socioeconomic theories to physical laws, they have done it, more often than not, by coupling their thoughts to Newtonian mechanics. Adam Smith (1723–90), who grew up when the mechanistic worldview was sweeping western civilization, formulated the theory that free markets are guided by an Invisible Hand, which keeps societies on track by ensuring that they produce the goods and services they need. Smith's Invisible Hand would act as a balancing agent, keeping supply and demand in equilibrium, thus stabilizing prices. As the demand for some economic item increased, entrepreneurs and producers would promptly rush to fill the supply gap, causing its price to gravitate in the proper direction.

But Adam Smith neglected the fact that this juggling act cannot be performed without increases in entropy. He also imagined that the balancing mechanism would adjust supply and demand instantaneously. "The equilibrium price-auction model is often taught in the classroom as if instantaneous adjustments occur, but technically the model is silent as to how long it takes markets to clear," writes economist Lester Thurow of MIT.[2]

For example, factories take a considerable amount of time to build. While a product can be sold promptly at less than the expected selling price if demand suddenly dwindles, the situation is different for the huge factories required nowadays to generate energy. It takes years to complete a nuclear power plant. Who knows what the demand for electricity will be then?

In addition, these long projects often experience cost overruns, which render them insolvent or economically unfeasible to complete. The Washington Public Power Supply System, more widely known by its satiric nickname Whoops, is a good illustration. Starting in the early 1970s, Whoops borrowed more than $8 billion for the construction of five nuclear generating plants in Washington State. As the project progressed, however, it amassed a record of "astronomical cost overruns."[3] By 1982, recession had sapped the demand for energy and the region needed no additional capacity. In 1983, Whoops defaulted on $2.25-billion bonds, shaking financial markets.

Because Adam Smith believed fervently in the Invisible Hand, he was against the idea of the government interfering with the inner workings of the economy. The free-market environment, if left alone, is supposed to make the system grow, and the nation should witness steady increases in its wealth. In Adam Smith's economic ideology, division of labor—specialization—plays an important role in achieving growth in production. However, he felt that division of labor can improve the rate of production—productivity—only to a point. From then on, the mechanism that takes over is "capital," mainly in the form of machinery and equipment.

According to this economic scheme, a nation can go on accumulating wealth by introducing more and more machinery to its stock of capital. Even Smith envisioned limits to economic progress as nations pushed economic growth to the natural limits of soil and climate. From his vantage point, however, an economic "heat death" was in such a distant future that for all practical purposes his recipe for economic growth should not be affected.

Is Adam Smith's economic growth prescription flawed? Yes, said Karl Marx (1818–83), who foresaw nothing but tension and friction between the workers and those who accumulate machinery and equipment. As firms amass more and more capital, they would get bigger and bigger, squeezing out smaller businesses in the process. With each economic crisis—which according to Marx is inherent in the capitalist system—some small businesses would go bankrupt, inviting bigger ones to buy their assets.

With the passage of time, the surviving businesses would get larger and larger, and would attempt to take control of the labor force, thus creating a class struggle between a small group of capitalist magnates and a large mass of embittered workers. This class conflict would gradually intensify, finally bringing the imposing economic citadel down to ruin. By the turn of the twentieth century, the capitalist system had not proved itself as unstable and chaotic as Karl Marx had envisioned, nor was it running as friction-free as Adam Smith had predicted. Then came the Great Depression, which caught everybody by surprise.

The free-market system was supposed to make the economic engine move forward smoothly. This time, however, the economic system would just not move. The consequences were devastating. In the United States, one fourth of the working force became idle. Over a million urban families found the mortgages on their homes foreclosed. When banks closed, 9 million savings accounts were lost; many banks never reopened. "Against this terrible reality of joblessness and loss of income, the economics profession, like the business world or government advisers, had nothing to offer," write economists Robert Heilbroner and Lester Thurow. "Fundamentally, economists were as perplexed at the behavior of the economy as were the American people themselves."[4]

Nobody knew how to fix the worn-out economic engine and get it rolling again, because nobody could pinpoint the cause of its extreme sluggishness. Amidst this dismay and helplessness came the great repairman John Maynard Keynes (1883–1946). Keynes's economic theory revolves around the notion that in the capitalist system the overall level of economic activity is determined by the willingness of its entrepreneurs to make capital investments. Unfortunately, said Keynes, this willingness is occasionally hindered by circumstances that make capital accumulation extremely difficult or impossible. When that happens, the economic system ceases to grow, begins to stagnate and decay, and ends up in prolonged economic depression despite the ample presence of idle workers and unused machinery.

The Great Depression convinced economists and government policymakers that Adam Smith's Invisible Hand was either imaginary or had died of exhaustion trying to equalize supply and demand. Since the Invisible Hand had reached its "maximum entropy" and had vanished, the capitalist system was left without a mechanism for revitalizing the system whenever it went into a state of depression. Keynes diagnosed that in Smith's scheme of laissez-faire, once the economic system reaches the stagnate state, it has no self-correcting mechanism that could bootstrap itself out of the doldrums into a period of renewed growth. Therefore, the economic system had no choice but to replace the Invisible Hand by a visible and concrete entity. That entity, according to Keynes, was the government.

In Keynes's scheme of things, whenever the economic system loses its vitality, government should act as the central stimulating agent for its rejuvenation. His rationale is simple: Since consumers and businesses are in a state of exhaustion and are incapacitated, they obviously lack the energy to spend money and make the economy roll. Therefore, a massive infusion of energy is required to revive the comatose economy. According to Keynes, it is the role of the government to perform that function in the form of increased expenditures.

The Great Depression demonstrated the falseness of yet another classical economic theory—Say's Law of Markets. This economic theory, named after the eighteenth-century French economist and writer Jean Baptiste Say, states that overproduction is impossible by its very nature. If goods are left unsold, prices fall to the level where everything is eventually sold. In addition, the existence of unsold goods means that consumption is declining and savings are rising. The increase in savings, in turn, will pull interest rates down and push investment up. Consequently, any slack in consumption is automatically offset by investment. In classical theory, the cyclical ups and downs of an economy are phases in the process of adjustment that always gravitates toward full employment. Thus no serious economic crisis can occur unless the government or the private sector interferes with the free market. Today, Say's law is expressed succinctly as "supply creates its own demand."[5]

As Keynes saw it, the Great Depression demonstrated that Say's law does not work. Prices did not fall very rapidly and savings failed to generate offsetting investment. He argued that a shortage of buying power could grow progressively worse as consumers reduce expenditures and businesses respond by curtailing production and investment. The economy could settle into a depressed level with high unemployment. To avoid such calamity, Keynes argued, the government should prop up demand through expenditures. In short, Keynes's answer to Say was, "demand creates its own supply."

Classical supply-siders are convinced, however, that Say was right and Keynes was wrong. They argue that the Great Depression was not caused by the inherent instability of the economy but rather by an ill-conceived series of government interventions that both accelerated the downturn and hindered the recovery. Such actions included a series of tax hikes, the tariff wars of the 1930s, a huge drop in the money supply, and anti-supply measures such as the deliberate destruction of agricultural products.[6]

Today's supply-siders and Keynesians seem to agree on one principle: that the government should intervene in the economic affairs of the nation and should stimulate the economy whenever it sees fit, in an effort to "control" the thermodynamic behavior of the economic system. In the late 1970s and early 1980s, when the U.S. economy was sluggish, with high unemployment and high inflation, modern supply-siders began a massive campaign to convince policymakers to change their prescription from demand-side to supply-side medicine; that is, to stimulate supply rather than demand, in conformity with the more fundamental Say's law.

The government, led by President Reagan, responded to the appeal of the supply-siders and began a massive "reform" in U.S. economic policy on two major fronts. The government gave businesses huge tax cuts and tax incentives to encourage investment and capital formation—thus stimulating the supply of goods. It also gave individuals large cuts in their personal tax rates to induce increased work effort—thus stimulating the supply of labor.

What happens to the federal deficit and the national debt? Some economists, including Nobel laureate Milton Friedman, argued that tax cuts and tax incentives do not necessarily increase the national debt; they may even reduce it. In an article entitled "Painless Revenue," Friedman maintained that the proposed tax cuts "would simultaneously raise the tax receipts of the Treasury and lower the cost of taxes to the taxpayers—a veritable free lunch."[7] The basic idea is that as the supply of capital, goods, and labor increases and the country produces more, the economy will grow substantially; consequently, the government will collect more money than it gave away in tax cuts.

Did the government's wish come to fruition? It did not. The Federal deficit took off, but in the opposite direction than envisioned by the government and its economists—the deficit increased massively and rapidly.

It took the government a few years to ascertain that the supply-side steering mechanism had gone awry; consumption had risen substantially and the national debt had skyrocketed—uncontrollably. The government was then forced to introduce yet another cybernetic process in an attempt to control its own behavior. It passed the Gramm-Rudman-Hollings law, designed to "automatically control" budget deficits by triggering automatic spending cuts if budget targets were not met.

Economists present their economic laws and views without mentioning any law of Nature. Consequently, whenever they put forward a new economic ideology or policy, they do not account for the effects of the Laws of Thermodynamics. In essence, they view the economic system as a purely mechanical system that can be controlled at will without any entropy increases. Understandably, their prescriptions for economic ailments are enticing because they appear to be virtually painless therapies with no discernible side effects. But every controlling mechanism requires energy to operate and generates entropy while in operation. As time goes on, the disorders created by the controlling mechanisms accumulate and become increasingly apparent.

For instance, when supply-side economics was being administered, the country was told that it would lift the economic tide and make all boats rise with it, thus benefiting everyone in the process. However, it takes energy to change the course of economic events, and eventually we must pay for it. In this case, the federal government boat that carries the nation's debt also rose with the tide, absorbing a huge amount of new debt in the process. Consequently, the nation as a whole must work harder to keep this heavy boat from sinking.

When a government intervenes in the economic activities of a nation in an attempt to control its behavior, it creates turbulence—entropy—in the socioeconomic system. While some boats benefit from the disorder, others begin to shake violently, and some even sink: while some people and businesses benefit, others have to pay for the entropy generated by the government's actions. By its socioeconomic engineering activities, the government divides the nation into groups—those who benefit from its actions and those who have to pay for them, directly or indirectly. The bigger the role of a government in the economic affairs of a nation, the bigger its contribution to the nation's disintegration into special interest groups.

The Economics of Computers and Technology

We are constantly reminded that computers—and technology in general—are inflation fighters. According to technology enthusiasts, advances in technology lead to efficiencies that, in turn, bring prices down. An example given is the sharp decrease in the price of computer hardware. In the words of futurist Alvin Toffler: "The speed with which computers have spread is so well known it hardly needs elaboration. Costs have dropped so sharply and capacity has risen so spectacularly that, according to *Computerworld* magazine, 'If the auto industry had done what the computer industry has done in the last 30 years, a Rolls-Royce would cost $2.50 and get 2,000,000 miles to the gallon.'"[8]

If we look inside today's automobiles, we can see one reason they have not followed the cost curve of the semiconductor. They have been invaded by computers, microchips, and semiconductors; their entropy has increased appreciably. Conservatively speaking, today's high-tech automobiles contain a few hundred dollars worth of semiconductors and computer-related technologies, all contributing to the price of the automobile, making it impossible to be sold for relatively few dollars. Moreover, the cost of a car due to semiconductors is rising as more and more exotic high-tech gadgetries are added. The car is gradually becoming a mobile computer and communication system.

Thermodynamics tells us to look at the total situation rather than a small portion of the equation. There is no question that the price of computer computations has plummeted since 1957. However, if we called our federal, state, and local government officials and asked them whether their overall computing expenditures have declined as dramatically as the price of semiconductors and computations per second, we would be surprised to learn that not only have their data processing costs not come down, they have gone up. Banks, insurance companies, utilities, hospitals, colleges, and other institutions will all tell us the same thing.

While it is true that the price of individual computer hardware has nose-dived, the total of all computer-related expenditures incurred by users—the most important number—has gone up as software costs, programmers' wages, training, and other computer-connected expenses have risen continually. In addition, what matters more to individuals and societies is not whether the price of a commodity has gone up or down, but whether they can function without that commodity. If they cannot function without it while previously they could, they will undoubtedly feel economic pressures from it although the price of the commodity has come down.

For example, 20 years ago college students were not required to have a personal computer. Today, parents of college students are feeling additional financial pressure because some universities have begun demanding that students own or rent a personal computer. And computers have also invaded high schools, junior high schools, and almost all institutions of learning, putting more pressure on parents. That a computer that now costs $2,000 used to cost a few million dollars 20–30 years ago is of no consequence or interest to today's parents. What matters is that they now have to provide computing power to their children in addition to food, clothing, and shelter.

Computer expenses—generated by the computer's entropy—are creating a socioeconomic problem for governments, whose role in recent years has been, at least in theory, to minimize inequities among individuals. Governments find this task even harder now, since there exists in the marketplace yet another machine that has to be distributed evenly among the citizenry to achieve a semblance of equality. Rich districts have the resources to provide computers to their schools and to children at home. How about poor districts? In 1983, the *Wall Street Journal* voiced a concern shared by many educators, that "computers, potentially a great equalizer, may instead widen the gap between the rich and the poor."[9]

If we examine what computers have replaced and are attempting to replace, it becomes clear why they are not deflationary. Before the arrival of the computer, engineers did their work with the aid of a slide rule that costs only a few dollars. Today's engineers cannot function without the support of expensive computers costing tens of thousands of dollars. The slide rule required no maintenance, while computers break down and require costly maintenance. One slide rule would carry an engineer through an entire career. This is not so with computers. Because of the high rate of obsolescence, engineers have to replace computers every few years. Moreover, what engineers design and build have computers in them as well, leading to products with the same high-entropic characteristics.

The overhead of other professionals, such as writers, doctors, lawyers, and accountants, has also increased as the computer has invaded offices. Because microchips and computers are everywhere, everybody's overhead is now higher than before. We all pay our dues to computers when we buy food, clothing, or a house; read a book, magazine, or a newspaper; fill up our car with gasoline; go to the doctor; or send our kids to school or college. The computer is included in the price of everything we do.

Computers, and technology in general, also have hidden costs, which technology enthusiasts forget to account for. Consider the so-called Y2K computer bug. Companies and government agencies in the United States spent an estimated $150 billion repairing the problem, which is more than the gross domestic product of many nations.[10]

The costly Y2K bug was a generic computer bug. There are countless other bugs specific to a given version of a particular operating system, software product, application software. These cost much time and money to patch and correct on a daily basis. They are a drag on productivity, which rarely shows up in statistics. They are also a constant source of aggravation and stress to programmers and users alike.

The idea that technology overpowers the forces of inflation is entrenched in many people's minds, including policymakers. In his 1985 State of the Union address, President Reagan enthusiastically told the nation how technological advances bring costs down. "In the late 1950s, workers at the AT&T semiconductor plant in Pennsylvania produced five transistors a day for $7.50 apiece. They now produce over a million for less than a penny apiece. New laser techniques could revolutionize heart bypass surgery, cut diagnosis time for viruses linked to cancer from weeks to minutes, reduce hospital costs dramatically, and hold out new promise for saving human lives."[11]

If technology is indeed a deflationary force, the government should certainly have seen some indication by now, since it is a major consumer of high technology. The U.S. government's cost of doing business—from computing to military hardware procurement to Medicare payments—should have shrunk to practically nothing. That is not the case. The reason lies in the fact that technology increases the complexity—entropy—of the thermodynamic system, which in turn increases the cost of living and doing business. This is true for governments, businesses, and individuals. A single high-tech bomber costs more than $500 million. In medicine, technological advances are enabling us to perform increasingly complex tests. Magnetic resonance imaging (MRI) can help detect tumors in the brain, lungs, respiratory tract, and liver, but an MRI facility costs a few million dollars.[12] Moreover, Robert Heilbroner and Lester Thurow point out, *"technology has brought a need for public supervision. An impressive amount of government effort goes into the regulation of problem-creating technologies."*[13]

Not only is high-tech machinery expensive to buy and install, it is also expensive to maintain. Machines cannot operate at zero entropy. They break down and wear out. With the passage of time, maintenance costs escalate as the internal entropies of the machines increase, until they become so prohibitively high that we are forced to discard them and replace them with more "advanced" models.

Today, we feel economic pressures from machines not only because they are more complex, but also because they have a much shorter lifetime than their predecessors. No sooner is a new military, medical, or computer technology introduced than the next technology is already conceived or designed. This high-entropic situation puts intense financial pressure on the working population, since they have to work harder and harder to subsidize, produce, consume, and replenish an ever-increasing number of complicated machines that are disintegrating at an increasingly faster rate. Moreover, because each technology brings a corresponding stock of new knowledge, whenever we make a given technology obsolete, we throw away the knowledge that people have acquired to design, develop, use, and maintain that particular technology. Consequently, we feel economic pressures from high rates of technological obsolescence as we strive to replenish a rapidly dissipating knowledge base.

In a high-entropic environment where complexity and technological obsolescence reign, R&D, capital, and education become expensive. "To train tomorrow's leaders, you need facilities with all kinds of sophisticated technological systems, systems that will be outmoded almost as soon as they're built," explained a senior vice president at Cornell University.[14]

Thanks to technological advances, we are surrounded with all kinds of machinery that allow us to do things we were never able to do before. The more things we do, the more we use energy and matter, and the more entropy we generate. Increases in entropy mean increases in the costs of doing business with Nature, which in turn effect increases in the costs of living. As entropy increases, we experience the degradation of the quality of energy-matter, the deterioration of the environment, and the dissipation of capital and knowledge. We definitely feel their effects on our economic well-being; we feel the squeeze from the Second Law.

The Concept of Environmental Externalities in Economics

Our failure to measure environmental externalities is a kind of economic blindness, and its consequences can be staggering. . . . Many U.S. policy-makers seem content to leave the environmental consequences of our economic choices in the large waste-basket of economic theory labeled externalities.[15]
—Al Gore

Given the record of this century, an extraterrestrial observer might conclude that conversion of raw materials to wastes—often toxic ones—is the real purpose of human economic activity. . . . And the effort to make waste disappear— by burying it, burning it, or dumping it in the ocean—has generated greenhouse gases, dioxin, toxic leakage, and other threats to environmental and human health.[16]
—Gary Gardner and Payal Sampat

In recent years, economists have reluctantly added a new variable in their economic thinking to account for the side effects induced by the production of goods. They have labeled the parameter "external-ities" (normally used in the plural due to its multiple effects) be-cause it usually affects, costwise, people other than those who are directly buying, selling, or using the goods in question.

When farmers produce food, economists call the product "goods." When the pesticides and fertilizers used in the production of these goods leak out of the farmland and pollute the nation's wa-ter supplies, economists call the pollution "bads," or environmental externalities, because the price of cleaning up the environment is not included directly in the price of the potatoes and apples we buy.

When nuclear reactors produce electricity, we are gratified be-cause we put electricity to such uses as washing and drying dishes and clothes. But when the nuclear reactions generate highly radioac-tive by-products, we are annoyed because the nuclear wastes are dangerous to our health. Economists call these unwanted nuclear wastes externalities, because most of the cost associated with stor-ing, regulating, and transporting them are not added directly into the cost of electricity. The Nuclear Regulatory Commission does not send a bill to the utilities for the cost of its bureaucracy.

While we enjoy using the products made in our factories, the generated smoke imposes additional costs in medical and cleaning bills on local households whether or not they use any of the fac-tory's output. "Externalities bring us to one of the most vexing and sometimes dangerous problems in our economic system—control-ling pollution," note economists Heilbroner and Thurow.[17]

Pollution, from an economic point of view, is the production of wastes, dirt, noise, and other things we do not want. It is an inevi-table consequence of producing things we do want. For example, we do want steel and cement, but we do not want the smoke pro-duced by the output process. We do want mechanical energy from heat engines, but we do not want the released heat, which we call—not quite rightly—"thermal pollution."

"The basic reason that externalities exist," say Heilbroner and Thurow, "is technological: we do not know how to produce many goods cleanly, i.e., without wastes and noxious by-products."[18] This explanation is not satisfactory. The truly fundamental reason why externalities exist is scientific, not technological. The Second Law of Thermodynamics states that it is impossible to perform energy transformations—which are required to produce things—without an increase in the entropy of the thermodynamic system. Consequently, we cannot produce goods or energy without also generating some externalities.

Economists' externalities are Nature's entropy. Since the middle of nineteenth century, we have known that all processes increase entropy. Yet only recently have economists begun talking about externalities—because only recently have humans become a high-entropic creature, generating massive amounts of entropy.

In earlier times, people used the Sun's rays to dry clothes; the process generated hardly any entropy. Today, we produce electricity instead and manufacture dryers to perform the same function; in the process, we generate additional entropy. Before, people ate unprocessed food. Today, we consume canned food, frozen TV dinners, and other packaged foods that generate large amounts of waste products—externalities. Before, people produced food in the amounts that the earth's soil allowed. Today, we use synthetic fertilizers to increase yields, but the excess nitrates are swept away, polluting our surface waters.

Externalities have become a major variable in industrial societies. As humans advanced technologically, we became a major producer of waste products that through the years have gradually accumulated to the point where we can no longer ignore their existence. When we began to develop nuclear energy some 45 years ago, we did not have any radioactive by-products to worry about. Since then, the United States alone has generated thousands of metric tons of "hot" radioactive debris. In the words of Sir Crispin Tickell, former U.K. Ambassador to the United Nations: "The fact that every year there is waste being produced that will take the next three ice ages and beyond to become harmless is something that has deeply impressed the imagination."[19] The cost of disposal impresses the pocketbook as well. Just to get an idea, the cost estimates of cleaning up the accumulated radioactive waste products at U.S. nuclear-weapons plants have soared to more than $100 billion.[20]

The costs of taking care of such externalities as dioxin, arsenic, and toxic incinerator ash are also mounting. "Depending on how you measure it, cleaning up the country's hazardous wastes is now a $35 billion industry," wrote *Forbes* magazine in 1987.[21] In 1998, U.S. industries produced 200 million tons of hazardous waste, much of it dispersed by uncontrolled disposal into the environment.[22]

Year after year, hazardous wastes keep piling up. Landfills are reaching capacity or becoming hazardous, and are being shut down. The United States has now some 40,000 locations that have been listed as hazardous waste "Superfund" sites. According to the Environmental Protection Agency, the cleanup of just 1,400 priority sites will cost more than $30 billion.[23]

Our modern economic life is also putting a strain on the air we breathe. Air pollution is particularly hazardous to children's health.[24] Los Angeles, Mexico City, and Santiago are among the cities suffering most from dense toxic smog.[25] In Chungking, China, when the fog and air pollution are at their thickest, locals say, "if you stretch your hand out in front of your face, you cannot see your fingers."[26]

Rivers are also being affected. More than half the world's major rivers are polluted or are going dry. Among the most stressed are the Colorado River in the United States, China's Yellow River, the Nile River in Africa, Russia's Volga River Basin, and the Ganges River in South Asia. In 1998, 25 million people fled their homes "because of the depletion, pollution, degrading and poisoning of river basins, outnumbering the war-related refugees for the first time in history."[27]

Environmental externalities are also generating friction among nations, as the entropy produced within the thermodynamic borders of one nation is crossing state lines and causing economic hardship to other nations. A typical example is acid rain, a source of tension between Japan and China, between the Scandinavian states and the United Kingdom, and between Canada and the United States. Japan's acid rain problems have dramatically worsened from China's coal burning: more than a third of Japan's acid rain originates in China.[28] In Scandinavia, the acid rain problem is caused largely by sulfur and nitrogen oxides emissions from the United Kingdom and Europe.[29] In Sweden and Norway, fish have died from acid rain in at least 6,500 lakes.[30] Similarly, the high acidity in the rain in Canada, which is generated by smokestack industries in the United States, has wiped out fish in lakes in Ontario and has damaged forests.[31]

Canada has its own internal environmental problems, including acid rain, generated by industry. A dramatic example is the ore-smelting operation of Sudbury, Ontario, population 150,000. There, local industry exploits the rich earth's massive deposits of copper and nickel. Sudbury's smelting facilities emit 2,750 tons of sulfur dioxide a day into the air, about 37 pounds per resident. To reduce the local impacts of this dense air pollution, the largest smelter in Sudbury built a smokestack 1,250 feet high. The idea is to push sulfur dioxide far from the town. But it did not go far enough. "Several hundred square miles of habitat just downwind of the smelter have been literally obliterated," reports Richard Turco, Professor of Atmospheric Sciences at the University of California, Los Angeles, and Director of UCLA's Institute of the Environment. "Photographs of the area show barren ground devoid of life stretching to the horizon. One is reminded of images beamed from the surface of the moon. The lakes in this region are acidic and totally dead."[32]

Most of the sulfur emitted from Sudbury smelters drifts far downstream, as far as the northeast United States, blending with pollution from other sources before falling back as acid rain. This is a classic illustration of externalities, where the entropies generated by industrial processes and their associated environmental costs are passed on to others without any liability. The acid fallout kills aquatic life by direct action on fish and by depleting the aquatic food chain. Local emission standards cannot curb acid rain and other pollutants that originate hundreds of miles away. On many environmental issues, the nation-state is obsolete. Consequently, we need multinational cooperation.[33]

Although Nature is a contributor to sulfuric and nitric acids, the evidence is overwhelming that the levels of acid rain we are encountering come from human activities.[34] They come from industrial society's inexorably growing use of cheap, bountiful coal to produce electric power, and the burning of fossil fuels in general. Sulfur dioxide and oxides released into the atmosphere—through natural processes like volcanic eruptions, or human activities like the burning of fossil fuels—end up in cloud droplets. The acid molecules remain suspended for a while in cloud form. Acid fog is an important health issue. People who are sensitive to airborne particulates suffer inordinately from acid rain. The killer fogs of London acted through the respiratory track. In southern California, fog has been detected that is more acidic than the most acidic rainfall ever recorded.[35] Eventually the acid molecules are washed out of the air in rain or snow and return to the earth, where natural processes in the soil work to dilute or neutralize the acidity before it can do any more damage.

However, human activities are overloading Nature's cleansing cycle, creating a host of environmental problems. "Spruces, maples and pines in California and Appalachia absorb the acids through needles, leaves and roots, and are now suffering from what the Germans poetically call *Waldsterben*, forest death,"[36] reported *U.S. News & World Report*. Germans are all too familiar with the phenomenon. Beginning in the early 1980s, "forest death" decimated stands of silver fir, Norway spruce, and European beech in the black forest of Bavaria, and later throughout Germany.[37]

In *The Dying of the Trees*, conservationist Charles E. Little tells us that trees are dying not just in Germany, California, and Appalachia but everywhere in the world, especially in the United States. Trees are mortally afflicted even before they are cut from a range of human-caused maladies—from fatal ozone to acid rain. "We are almost certainly witnessing the accumulated consequences of some 150 years of headlong economic development and industrial expansion, with the most impressive of the impacts coming into play since the 1950s—the age of pollution."[38] While technology enthusiasts call the era we are living the age of information, observers of Nature call it the age of pollution—the age of entropy production—which includes information with all its disorders.

The overall cost of environmental externalities—from nuclear wastes, acid rain, and toxic wastes to air and water pollution—is staggering. Whichever way we turn, externalities have bulged to such a level that we can no longer sweep them under the rug. The increase in the disorder of our environment is Nature's way of taxing our economic activities. The economic system must recognize this fact and somehow account for the externalities—entropies—that our economic activities generate, but it rarely does.

There is a human side to the economic externalities, which cannot be told through historical anecdote, surveys, statistics, or mathematical equations. The real story can be found "in the mounting number of human dramas played out by ordinary people whose lives have been changed overnight," write Fred Setterberg and Lonny Shavelson of Pacific News Service.[39] For three years they listened to hundreds of people confide, rail, and worry that "the unfettered proliferation of toxic wastes had devastated their lives." They crisscrossed the United States talking with people "who believed that their families were imperiled by a pervasive menace that had been churned out into the environment by hazardous waste dumps, dioxin-spewing industrial chimneys, toxic waste incinerators, pesticide-spraying airplanes, home garden weed killers, apartment buildings saturated with formaldehyde, chemical food contaminants, leaking landfills, legal and illegal dumping, and industrial accidents."[40] It turned out their fears were justified; many were chronically sick from environmental problems.

In *Earth Under Siege: From Air Pollution to Global Change*, Richard Turco writes:

> In some ways, human civilization has been treating the global environment like a bottomless trash pit. Cities have filled up pristine canyons and valleys with rubbish. Countless tons of human garbage and sewage have been dumped into rivers and off coastlines. Humans are the only organisms that generate useless rubbish, including not only indestructible solid waste, but also toxic and radioactive materials that poison and despoil the air, soil, and water. . . . Each of the major biogeochemical cycles that hold our environment together is under the relentless assault of human activities. Each cycle is beginning to show major changes directly traceable to human activities.[41]

Turco supports his conclusion with massive documentation and scientific facts. When planet Earth is under siege, humanity is under siege.

As it turns out, the entire biological world is under siege. There is another kind of externality that human economic processes are causing—the mass extinction of species. Economists do not recognize the value of biological diversity for maintaining the health of the planet, but biologists do. To some biologists, such as E. O. Wilson, the loss of biodiversity is an important aspect of the global ecological crisis.[42] We find ourselves in the midst of the greatest decimation of plant and animal life in 65 million years, a global evolutionary convulsion with few parallels in the entire history of life spanning 3 billion years.[43]

Extinction is a rare but natural phenomenon. The "background" rate of extinction lies between 1 and 10 species a year. The current extinction rate is at least 1,000 species a year, an indication that planet Earth is sick—and "we are the reason for it," as John Tuxill and Chris Bright of Worldwatch Institute point out.[44] The leading culprits in the loss of biodiversity are a set of interrelated factors all pointing to human activities. Through our numbers—about 6 billion—and resource use, humans have become a dominant force in the increase of Earth's entropy, its ecological disorder.[45] From deforestation, urbanization, and the alteration of the biogeochemical cycles to chemical pollution, acid rain, and the alteration of the global climate, humans are wrecking havoc to the ecosystem of planet Earth.

It is crucial to recognize the scientific fact that we cannot possibly eliminate environmental externalities by any technological means whatsoever. In general, environmental externalities are solved in two ways: either by substituting one thermodynamic process for another (substituting a "clean" process for a "polluting" one), or by bringing in a second process to counteract the disorders generated by the first one. In the case of the greenhouse gases emitted by our automobiles, Gregg Easterbrook wrote in *Newsweek*: "America burns more oil than any other nation. . . . We could be driving higher mileage cars, or switching to fuels such as ethanol that not only pollute less but subtract carbon dioxide during production, while growing as corn."[46] In the same article, however, Easterbrook wrote about water pollution: "Today just 9 percent of stream pollution comes from industry. Fully 65 percent is nonpoint, primarily from agriculture."[47]

Increases in corn production do not come about without increases in entropy. More corn production means more usage of fertilizers, pesticides, and water in general. Externalities like water pollution from fertilizers and pesticides would increase, and our topsoil would deteriorate at a faster rate. We should also not forget that agriculture—chemical or organic—is a major consumer of energy, which means that it is a major producer of entropy. As another article in *Newsweek* has pointed out, "if ethanol is made from corn, the oil that powers the tractors and the natural gas that produces the fertilizer to grow the crop may dirty rural air enough to offset gains in the cities."[48]

Another fuel that has been hailed as a "clean-burning fuel of the future" is methanol—wood alcohol. It can be produced from such diverse sources as wood, coal, natural gas, and biomass. Its greatest virtue is that it produces less smog-forming soot: 35–50 percent fewer hydrocarbons, and 30–40 percent fewer airborne toxins.[49] But as an advertisement by a major oil company remarks, "methanol as an automotive fuel could create as many new environmental problems as it would solve—that in the end we may have merely substituted one set of problems for another, at a very high cost."[50] What are the new problems? For one thing, the burning of methanol emits even higher levels of the noxious gas formaldehyde, up to eight times as much as gasoline exhaust.[51] What does this noxious gas do? "Formaldehyde," explains the ad, "is strongly reactive in producing ozone, the very pollutant that methanol use is supposed to reduce. And no one has demonstrated how to control formaldehyde emissions from methanol vehicles." How would we control formaldehyde emissions? By introducing yet another entropy-producing process.

As it turns out, methanol is far more toxic than gasoline. Ingesting even small amounts can cause blindness and even death. Moreover, methanol, like gasoline, releases heat-trapping carbon dioxide into the atmosphere when burned, although a bit less. Thus while it may make the air more breathable, it will not forestall the greenhouse effect.[52] Methanol also generates photochemical smog, though the fumes are sweeter-smelling than gasoline exhaust.[53]

Alcohol fuels—ethanol and methanol—have different chemical characteristics than gasoline. They are corrosive to ordinary steel and other materials. Brazil was the first to experiment, starting in the mid-1970s, with cars powered by ethyl alcohol made from sugar cane. Brazilians were thus the first to discover, through experience, the corrosive property of ethanol, which engineers call the "aggressiveness of the fuel." This forced automakers to apply a protective coating to all parts that come into contact with the alcohol. The additional processes mean, of course, more production of entropy and more expenses. In addition, alcohol's low volatility makes starting a car almost impossible in temperatures under 59 degrees Fahrenheit. Consequently, car makers had to design and install various cybernetic "cold start systems," thus adding more complexity to the car.[54]

Because of the aggressiveness of the fuel, ethanol-fueled vehicles do not last long. In Southern California, the Metropolitan Transportation Authority found out that the engines of ethanol-fueled buses, while providing cleaner emissions, "were notoriously fragile, . . . 'burning out' on average only one year into their expected 12-year life cycle," hardly a bargain.[55]

Another substitution of a clean energy for air-polluting fossil fuels comes from the nuclear industry. As an advertisement entitled "EARTH DAY, Preserving the Environment" states: "Every day is Earth Day with nuclear energy." Among the reasons given is the assertion that "Nuclear energy doesn't emit greenhouse gases. Because nuclear plants generate electricity cleanly, *every day* nuclear energy helps reduce greenhouse gas emissions from utilities by 20%." The ad also reminds us that "Nuclear energy helps reduce air pollution. *Every day*, by using nuclear-generated electricity Americans help reduce airborne pollutants by over 19,000 tons."[56] While it may be true that nuclear plants generate electricity "cleanly" and help reduce airborne pollutants, they do nevertheless generate, every day, a considerable amount of entropy—heat and long-lived, highly radioactive by-products.

We cannot eliminate environmental entropy by shuffling the disorder around, or by adding a second process to take care of the entropy generated by the first one. It is no secret that technological societies, including the United States, produce billions upon billions of pounds of toxic chemical wastes annually, and that landfills are filling up. In many instances, modern incinerators are used to take care of this environmental problem. But incineration is yet another process that generates its own entropy. *Time* magazine describes the situation: "Incinerators, burdensome investments for many communities, also have serious limitations: contaminant-laden ash residue itself requires a dump site."[57]

The sea odyssey of the freight *Khian Sea* brought forward the severity of the environmental problem produced by incinerators. For more than two years this freighter sailed around the world looking for a port that would accept its cargo—14,000 tons of toxic incinerator ash. Finally, the captain announced that he had unloaded the ash in a country he declined to name.[58] During this bizarre globe-circling trip, one of the hundreds of documented efforts to export toxics to faraway places, the name of the infamous ship had changed from *Khian Sea*, to *Felicia*, and finally to *Pelicano*.[59]

As toxic waste products pile up, industrial nations like the United States are running out of places to put them. This condition has been a catalyst for a new international commerce in hazardous waste. "Simply put, the world's wealthiest nations are exporting vast quantities of toxic waste [externalities]," wrote investigative reporter Bill Moyers. "One more burden for the poor of the world."[60]

The message from thermodynamics is clear: To reduce environmental externalities, we must reduce energy transformations and processes. Economists and policymakers have to recognize the fact that the production of externalities is not a technological issue but a scientific one emanating from the Second Law of Thermodynamics, which cannot be repealed.

Economics as an Applied Science

Page after page of professional economic journals are filled with mathematical formulas leading the reader from sets of more or less plausible but entirely arbitrary assumptions to precisely stated but irrelevant theoretical conclusions.[61]
—Wassily Leontief, *Nobel laureate in economics*

In his celebrated book *Economics*, Nobel laureate Paul A. Samuelson writes, "As if to commemorate the coming of age of political economy as a science, a new Nobel prize in economics was instituted in 1969."[62] Unquestionably, economics has become a mathematical discipline. Whether it has indeed become a science or an applied science remains debatable. Economists are not happy about the performance of their discipline. Some respected economists have been critical of their field, even to the point of suggesting that "economics is not only a dismal but a feeble science."[63]

Among the dissatisfied economists is Lester Thurow. In *Dangerous Currents: The State of Economics*, he has unkind words for theorists. He presents and then methodically demolishes several economic theories, both current and venerable, concluding that conventional economic theories just do not work in practice. He notes that economics textbooks are "moving toward narrower and narrower interpretations. The mathematical sophistication intensifies as an understanding of the real world diminishes."[64] To economists' chagrin, the world is not behaving the way it should. "No other discipline attempts to make the world act as it thinks the world should act," writes Thurow. Worse yet, when the stubborn facts do not match economic theories, "instead of adjusting theory to reality, reality is adjusted to theory."[65]

Milton Friedman is equally unhappy about economists' achievements, as his book's title, *Bright Promises, Dismal Performance: An Economist's Protest*, indicates.[66] Economist Eliot Janeway's review of Friedman's opus is entitled "Making Monkeys Out of Economists and Bureaucrats."[67]

The most discontented economist is probably Wassily Leontief, who was sufficiently upset on the dismal state of academic economics to write an open letter to *Science* magazine in 1982. He pointed out that economics has become highly mathematical and at the same time far removed from economic reality. A perusal of the contents of the *American Economic Review* spanning 10 years revealed that only 1 percent of the articles were analyses based on data gathered by the author's initiative. More than 50 percent of the studies were "mathematical models without any data." Leontief wrote: "Year after year economic theorists continue to produce scores of mathematical models and to explore in great detail their formal properties; and the econometricians fit algebraic functions of all possible shapes to essentially the same sets of data without being able to advance, in any perceptible way, a systematic understanding of the structure and the operations of a real economic system."[68]

Economists disagree more than ever on the causes of economic ills. In the area of solutions, the disagreement is even more intense. Economics Nobel laureate Robert E. Lucas, Jr., of the University of Chicago thinks that "economists should go on trying to discover how the world works," though he believes that "the search is likely to be long and arduous."[69]

The suggestion that economists should make an effort to learn the way the world works is a refreshing idea. Nicholas Georgescu-Roegen, who called himself an "unorthodox economist," looked outside the field of economics to the discipline of thermodynamics to discover how things work.[70] In *Energy and Economic Myths*, he brought to light a curious event in the history of economic thought. He points out that years after the mechanistic worldview had lost its supremacy in physics and its hold on the philosophical world, the founders of the neoclassical school of economics set out to build an economic science after the laws of mechanics—in the words of

Jevons, as *"the mechanics of utility and self-interest."* While economics has made great strides since then, there has been no deviation from the mechanistic epistemology of the forefathers of economics. A glaring proof is "the standard textbook representation of the economic process by a circular diagram, a pendulum movement between production and consumption within a completely closed system."[71]

The situation is the same with the analytical pieces that adorn the standard economic literature; they also reduce the economic process into a self-sustaining mechanical analogue. In this respect, Marxist economics is no different. In Marx's famous diagram of reproduction, the economic process is also depicted as a circular and self-sustaining affair.[72] Unquestionably, the mechanistic point of view is entrenched in economic thinking and theories.

Some economists have ventured beyond Newtonian mechanics. They have alluded to the fact that humans can neither create nor destroy energy. What Georgescu-Roegen found amazing was that "no one seems to have been struck by the question—so puzzling in the light of this law—'what then does the economic process do?' All that we find in the cardinal literature is an occasional remark that man can produce only utilities, a remark which actually accentuates the puzzle."[73]

The answer to the puzzle about what economic processes do can be found in the Second Law. Economic processes are no different than all other natural processes. While it is true that humans cannot create energy or matter, we can nevertheless create utilitarian products through energy-matter transformations. The economic process does not create or annihilate energy-matter; it only makes transformations. For example, it may take in natural resources as input, and transform them into goods and products.

Nature does not just sit back and watch us perform our energy-matter transformations, nor does she applaud our technological feats. Instead, she sends us her Second Law—Nature's relentless Internal Revenue Service—to collect her dues. In economic terms, the Second Law exacts a tax from all our economic processes by increasing the entropy—disorder—of our thermodynamic system. (See Figure 8.1.) No process eludes Nature's Second Law; her tax code contains no loopholes and no tax shelters.

To become an applied science, economics must recognize and work within the Laws of Nature. Economic thought cannot be ad-

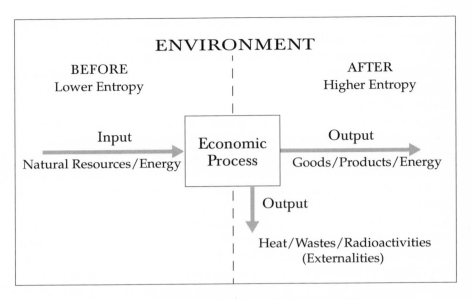

Figure 8.1 Thermodynamic View of Economic Processes

vanced only by criticizing and attacking failed economic theories, a popular activity in recent years. Eventually, economists must—as Robert Lucas has suggested—discover how the world works, and incorporate Nature's Laws into their economic theories. Only then can economics become an extension of the natural sciences.

As it turns out, economists are beginning to recognize the existence of the Laws of Thermodynamics. In the eleventh edition of *Economics*, Paul Samuelson acknowledges that economic laws "must respect those of nature and physics." And he singles out the Laws of Thermodynamics:

> You can't make a perpetual motion machine that will run by a dropped ball's bouncing back to *higher* than its point of release. That's a consequence of the first law of thermodynamics—which guarantees conservation (or constancy) of the total of energy.
>
> More subtle, but no less consequential for economics, is the second law of thermodynamics: It requires that the total of *entropy* (or "disorder") *irreversibly increases*, while the total of energy is remaining constant.[74]

The Second Law is indeed subtler and more consequential for economics than the First Law. What is more important, the Second Law has two sides. The side that has received more attention in the literature, including Samuelson's *Economics*, is the less fundamental one: the one that describes entropy as a measure of the unavailable energy in a thermodynamic system—one of the dictionary definitions of entropy. This side of the law is closely related to the concept of the dissipation of energy-matter. "The ball you dropped won't bounce back even to its previous height," writes Samuelson. "Some of its energy is dissipated in useless frictional heat."[75] It is this interpretation of the Second Law that leads us to conclude that economic processes—like all natural processes—degrade the quality of energy and matter by transforming them into less available forms. Consequently, with the passage of time, the availability of energy-matter decreases, creating economic hardships for humankind.

Economists attack this problem in many ways. In one approach, a common one, they turn the problem around 180 degrees—from physics back to economics—and solve it through economic principles: "No doubt many sincere persons believe that the world is actually 'running out of energy,'" notes economics writer James Dale Davidson, adding:

> Because energy and matter are two configurations of the same thing, we can never run short of either—as long as the universe exists. The only challenge we face is an economic challenge: finding and applying the cheapest form of energy for a given purpose. . . . We must merely face the economic challenge of paying for what we use. That is part of what Emerson called "the absolute balance of Give and Take, the doctrine that everything has its price." The higher that price, the more energy is available and the less energy is needed.[76]

Davidson is alluding to the First Law, the law of conservation of energy. His approach, used by many economists and policymakers, bypasses the laws of physics that govern energy and matter. If there are other laws of physics related to energy-matter, they become inconsequential to economics because economic principles automatically take control over them—through such things as the Give and Take doctrine, the laws of supply and demand, Adam Smith's Invisible Hand, or supply-side economics. In this approach, economists and policymakers do not need to learn Nature's Laws beyond a casual acquaintance with the First Law.

Economists attack the problem of the decreasing availability of energy in another way. They first acknowledge that we are depleting and using up "irreplaceable deposits of fossil fuels and pure ores."[77] Then they quickly start enumerating some of the weapons we have in our arsenal that can render the Second Law impotent and inconsequential. Nuclear energy—fission and fusion—ranks high on the list. Samuelson writes: "*Fission* or splitting of uranium atoms can . . . provide alternatives to the using up of oil, natural gas, coal, and other fossil fuels laid down over eons of geologic time. Safer because of its freedom from radioactivity as a by-product, controlled *fusion* of light atoms . . . can release even greater amounts of useful energy."[78] (But see Chapter 9 for a discussion of how fusion is not free of radioactivity.) Fusion is the darling of many economists. Whenever they want to emphasize the point that energy sources are practically inexhaustible, they invariably mention nuclear fusion.[79]

Economists must recognize that the Second Law is not merely about the availability of energy. If it were, it would deserve no more than an honorable mention, with a reminder that the Sun will be shining for another 5 billion years, and the case would be closed. However, the Second Law has another side—far more basic—which economists cannot ignore, the side that governs all economic processes.

If we examine the other side of the Second Law, the picture changes diametrically. As Max Planck repeatedly and expressly pointed out, the other side of the Second Law has a broader meaning, the side that states that in all natural processes entropy always increases—even when no energy transformation has taken place.

For example, when the supertanker *Exxon Valdez* plowed into a rocky reef off Valdez, Alaska, a natural process with enormous economic consequences took place without the occurrence of any noticeable transformation of energy. The accident spilled some 35,000 tons of oil into the ocean. The oil that gushed from the tanker immediately began to spread like a black, sticky wave through the bountiful waters of Prince William Sound. In a week, through the process of diffusion, the oil spread over some 900 square miles, coating thousands of marine mammals, birds, and fish, threatening hatcheries and parklands.[80]

Before and after the mishap the same quantity of oil remained in our thermodynamic system, yet the entropy of our environment increased substantially. The event set in motion a host of economic activities: lawyers were called to initiate lawsuits; cleanup crews were set up to remove the oil from the waters, rocks, beaches, and coastlines; marine biologists and volunteers came to rescue the mammals and the birds; oil prices fluctuated and gasoline retailers as far away as Chicago raised prices at the pump; Alaska's $750 million-a-year commercial fisheries took a beating.[81]

Among the known effects of the oil spill was "a huge loss of wildlife"—maybe as many as 5,000 sea otters, 300 harbor seals, 22 killer whales, 150 bald eagles, and an estimated quarter-million waterfowl and other birds.[82] After a decade, the injured ecosystem has not recovered; there are still signs of trouble among sea otters and other species, and continuing presence of oil residue. "The ecosystem that is there today is not the ecosystem that was there before the spill," notes *National Geographic*, partly because of the effects of the spill and partly because of natural change. The social effects of the spill are another issue: the disruption of human lives, manifested by busted marriages and domestic violence, is also not recoverable.[83]

As it turns out, a megaspill like the Valdez is not unique. The larger wrecks such as the *Torrey Canyon* (120,000 tons) in 1967 and the *Amoco Cadiz* (220,00 tons) in 1978 had already shown that the problem is indeed structural and worldwide.[84] About 3 billion tons of crude oil or petroleum products are shipped around the world annually. "In the process," writes economics professor Ravi Batra of Southern Methodist University, "2 million tons slip into the marine environment from routine tanker operations like tank cleaning and ballasting and oil spills from tankers and platforms."[85]

Economists must look at both sides of the Second Law. If they looked only at the side that relates entropy with the amount of unavailable energy, they would conclude that the Second Law could be dealt with—almost indefinitely—by making more and more energy available for economic use. And the more energy we make available, the more irrelevant the Second Law becomes. But what happens is just the opposite. As we increase energy availability and transformations, we simultaneously increase entropy production. Consequently, we feel the impact of the Second Law even more acutely, because our taxes to the law become more pronounced.

If economists would like to see economic theories reflect reality, instead of adjusting reality to economic theories, they need to gradually overhaul their theories and textbooks to conform to the Laws of Thermodynamics. Since Georgescu-Roegen has already started this work (*The Entropy Law and the Economic Process*—1971), they need not begin from scratch.[86] In some cases, economists may have to come up with a new economic state variable, while in other cases they will recognize that the variable has already been discovered. For example, economics already has a parameter—a fundamental one—that is practically synonymous to the concept of entropy. It is the externalities described earlier.

As economist Herman Daly of the University of Maryland points out, if economics adopted the thermodynamic view and incorporated into the "standard text" Georgescu's entropic flow of economic processes, "the standard text would require so many revisions as to make the text no longer standard, even though many chapters would still look familiar." Daly, who studied under Georgescu-Roegen at Vanderbilt University, adds: "Perhaps this is why such changes are glacially slow in coming. One does not expect fundamental change to occur overnight. But twenty-five years is a reasonable time over which to hope for progress. What is the matter with our discipline?"[87]

There is another issue besides the laborious task of revising the standard economic text. It is troublesome, if not embarrassing, for any discipline to admit that its theories are fundamentally flawed, more so for economics because its best receive Nobel prizes. Even in the natural sciences, fundamental change in theories takes time. It took half a century for the caloric theory to be overthrown, because its adherents, who were influential, vigorously fought to preserve the theory. Finally, the theory was brought down not by the members of the establishment but by three outsiders. According to Daly, Georgescu's latter years were marred with bitterness because the economics academia had ignored his theories, which were based on the Laws of Thermodynamics. "So great was his bitterness that he even cut relations with those who most valued his contribution," writes Daly. "But none of that diminishes the great importance of his lifework, for which ecological economists must be especially grateful."[88]

Similarly, Ludwig Boltzmann became depressed at the end of the nineteenth century fighting the powerful physics establishment for the acceptance of his theory, and eventually committed suicide. In his autobiography, Max Planck tells that it was clear to him that Boltzmann's theory would eventually triumph, adding: "This experience gave me also an opportunity to learn a fact—a remarkable one, in my opinion: A new scientific truth does not triumph by convincing its opponents and making them see the light, but rather because its opponents eventually die, and a new generation grows up that is familiar with it."[89] In the case of economics, there is some urgency for economists to assimilate the Laws of Thermodynamics into their theories, because economic theories are driving our world into dangerous levels of entropy production. Even the usually reserved London *Financial Times* ran an article with the headline "Save Planet Earth from Economists."[90]

The concept of entropy can be used to measure the overall quality of a nation's gross national product (GNP), its total economic output of goods and services. Entropy, which measures the disorder of a thermodynamic system, also measures the state of a nation's economic situation. Economists can analyze the economic health of a nation by examining the nature of the various economic activities that contribute to the GNP.

For example, if the economic system is building lots of prisons because criminal activities have increased, the GNP shows growth. If the participants in the economic system are buying guns in droves, the GNP also shows growth. In neither case does the GNP tell us the country is becoming a dangerous place to live. If sales of security devices increase, the GNP again shows economic growth, but fails to recognize that the social fabric of the nation has deteriorated to the point where too many people feel the need to install a security system in their homes, apartments, automobiles, computers, and businesses. If consumption of anti-depressants increases, the GNP shows economic growth and power, yet its participants are in a state of depression. If toxic and radioactive waste products increase, the GNP shows growth, but fails to recognize that the landscape is in the process of turning into a hazardous dump.

More and more economists are recognizing that neither GNP nor gross domestic product (GDP) is a good indicator of economic success. As economics Nobel laureate Amartya Sen has pointed out, the GNP neglects critical socioeconomic factors such as well-being.[91] Homelessness is often found near multimillion-dollar homes, famine often coexists with plentiful food, and high starvation rates are common in wealthy nations. In 1998, while the U.S. GDP grew a healthy 4 percent, requests for emergency shelter in the country rose 11 percent.[92] On any given night, some 600,000 Americans are homeless.[93] The mayor of Burlington, Vermont, Peter Clavelle, describes the state of affairs this way: "The economists tell us that we're living in an era of unprecedented prosperity, yet the food banks and homeless shelters tell us that the lines are getting longer."[94]

Other economists emphasize that GNP does not account for the depletion of natural resources and the environmental entropies produced by the economic activities. In recent years, China's economy has been growing at about 10 percent a year, a remarkable achievement. However, the official *China Daily* estimates that the yearly cost of environmental degradation is 7 percent of the GDP. Vaclav Smil, a geographer at the University of Manitoba and an expert on China's environment, has calculated the cost "at no less than 10 to 15 percent of GDP."[95] If this is the case, then the much-heralded expansion of China's economy is being canceled out by associated environmental degradation and entropy. Thermodynamics tells us to look at the total situation. In economic accountings, we cannot disregard the environmental entropies produced by economic processes.

Chapter 9

Why Things Look So Good on the Horizon–Until We Get There

Things were good once, and they'll be good again; the only trouble with the world is that it's right now.[1]
—Jack Webb

Why Great Expectations Turn to Disillusionments

"It began with such a promise," wrote *Time* magazine in 1984 about nuclear energy. The scientists and engineers were going to harness the tremendous energy available in the nucleus of the atom. "They would build nuclear power plants producing electricity so easily that it would be 'too cheap to meter.' At a time when technology promised an almost boundless potential for improving humankind, nuclear power seemed so modern."[2] As journalist David Deitz, a winner of the Pulitzer prize, proclaimed, "The day is gone when nations will fight for oil."[3]

A few decades later, however, the building of nuclear power plants has become extremely complicated, expensive, and full of dangers. The nuclear industry is having a difficult time producing electricity at any cost, let alone giving it away.[4]

It is becoming increasingly obvious that the construction of nuclear plants is more complicated than the prophets of nuclear power had led the public to believe. Complexities of the machinery have led to various nuclear incidents, such as the machinery malfunctions and operator errors that brought the reactor close to nuclear meltdown at the Three Mile Island plant in Pennsylvania.

In 1986, a more severe accident occurred at the Chernobyl's nuclear power plant in Ukraine. Two powerful explosions in the reactor released radioactive particles for nearly 10 days, which spread over a wide area in Europe and eventually across the whole northern hemisphere. The total radioactivity of the material released was estimated as 200 times the combined releases from the atomic bombs dropped on Hiroshima and Nagasaki during World War II. Within three months, 31 people died from acute radiation sickness; 375,000 people were permanently evacuated from the region; more than 50,000 square miles remain contaminated. Thyroid cancer has increased dramatically among children in Belarus and Ukraine.[5]

When technology gurus spoke of the enormous possibilities of nuclear energy, they did not mention, or account for, the entropy that nuclear reactions would generate over time. Some 45 years later we find ourselves amidst enormous quantities of highly dangerous, spent nuclear fuel rods that no one wants in their own backyard. These rods will remain radioactive for thousands of years. "A new generation accustomed to seeing the dark side of technology sometimes views nuclear power as the future that did not work," remarked *Time*.[6]

Consider what was said about the amazing automobile in the prestigious *Scientific American* in July 1899: "The improvement in city conditions by the general adoption of the motor car can hardly be overestimated. Streets clean, dustless and odorless, with light rubber-tired vehicles moving swiftly and noiselessly over their smooth expanse, would eliminate a greater part of the nervousness, distraction, and strain of modern metropolitan life."[7] No scientist or layperson would make such a statement today. Automobiles are poisoning the air, contributing to the greenhouse effect, and spoiling the quality of life with noise and strains.

These examples demonstrate a deep-rooted principle of Nature. Whether we are dealing with nuclear transformations, automobiles, or any other thermodynamic process, we should not forget that every machine, energy transformation, or event has associated with it some entropy. For every potato and steak we eat, we have some topsoil deterioration and environmental disorder; for every medicine we take, we experience some side effect; and for every chemical, car, refrigerator, computer, robot, and airplane we manufacture, use, and discard, we have some energy and ecological degradation. This is the essence of the Second Law of Thermodynamics. Various names are given to entropy. As noted in earlier chapters, economists call it externalities, ecologists call it waste products, physicians call it side effects, physicists and chemists call it disorder, environmentalists call it pollution, sociologists call it the dark side of progress, while historians call entropy the unintended consequences of technology.

In *Why Things Bite Back: Technology and the Revenge of Unintended Consequences*, Edward Tenner documents, sometimes humorously, "the unintended, ironic consequences of the mechanical, chemical, biological, and medical forms of ingenuity that have been hallmarks of the progressive, improvement-obsessed twentieth century." He concludes: "The technological dream of a self-correcting world is an illusion no less than John von Neumann's 1955 prediction of energy too cheap to meter by 1980."[8]

Whenever a new technology, ideology, economic theory, or scientific discovery comes about, it arouses great expectations among its believers and followers. Before its application, there is no entropy associated with the process, only promises of great returns. People raise their hopes in anticipation of unprecedented accomplishments yet to come. Then, as the ideology, technology, or scientific breakthrough is applied, the entropies of the processes begin to appear, creating new, unanticipated problems. As the disorders pile up, hopes diminish.

These cycles of hopeful visions followed by disenchantment have occurred in all areas where humans have ventured. Science and technology have been particularly vulnerable to these ups and downs. As physics Nobel laureate Val L. Fitch observes, "These shifts in attitude toward science illustrate that the country tends to operate in gigantic oscillations—cycles of enthusiasm followed by cycles of disillusionment."[9]

We have fairly recently experienced disappointments even in the formal or purely theoretical sciences. For example, when mathematicians began to develop mathematics on a large scale in the seventeenth and eighteenth centuries, many of them thought that they were constructing a perfect system. The awareness that mathematics did not offer certainty came about when mathematicians and logicians discovered serious contradictions in what was supposedly the perfected mathematics.

"Today there is no agreement among mathematicians on fundamental principles," declared mathematician Morris Kline. "So mathematics is not the universally accepted, precise body of knowledge that it was thought to be 100 years ago when scholars believed that it revealed the design of the universe. We don't believe that anymore. . . . Mathematics is a man-made, artificial subject. It is not *the truth*."[10]

Psychology also started with great expectations. Scientists were going to study the mind and come up with the appropriate processes and steering mechanisms to govern behavior. Thrust by Harvard's B. F. Skinner, psychologists began to develop an effective science of "behavioral engineering" with the hope of perfecting the technological expertise that would lead the individual to control his or her own conduct.[11]

Skinner's technique of behavioral engineering was put to test by two Skinnerians, Keller and Marian Breland, to find out if the science would work outside the laboratory, to determine if animal psychology could stand on its own feet as a genuine engineering discipline. In 1961, the Brelands published their findings in the form of a short, funny article entitled "The Misbehavior of Organisms," in a parody of Skinner's own major work, *The Behavior of Organisms*.[12] They used Skinner's operant conditioning technique to train 38 species, totaling over 6,000 individual animals.

The Brelands were successful up to a point. When they ventured further and further from the security of the Skinner box, they ran "afoul of a persistent pattern of discomforting failures." In their words: "These failures, although disconcertingly frequent and seemingly diverse, fall into a very interesting pattern. They all represent breakdowns of conditioned operant behavior."[13] The animals did not do what they had been conditioned to do, representing a "clear and utter failure of conditioning theory."[14]

Behavioral technologists, on the other hand, claim to be closer than ever to the long-sought goal of achieving a fully engineered human psychology. "I believe," wrote professor of psychology James V. McConnell, "that the day has come when we can combine sensory deprivation with drugs, hypnosis and astute manipulation of reward and punishment to gain almost absolute control over an individual's behavior. . . . I foresee the day when we could convert the worst criminal into a decent, respectable citizen in a matter of a few months—or perhaps even less time than that. . . . Today's behavioral psychologists are the architects and engineers of the Brave New World."[15] His article was written in 1970. The day when vicious criminals are transformed into docile, productive members of society through psychological engineering and force has not yet arrived. Crime is still rampant, especially in cities.

However, Dr. José M. R. Delgado foresaw the day when civilized humans would attain a level of "psychocivilization" adorned by a population "happier, less destructive, and better balanced than present man."[16] In 1969, Delgado detailed his vision in *Physical Control of the Mind: Toward a Psychocivilized Society*. He wrote, "We are facing a situation in which vast amounts of accumulated destructive power are at the disposal of brains which have not yet learned to be wise enough to solve economic conflicts and ideological antagonisms intelligently. The 'balance of terror' existing in the present world reflects the discrepancy between the awesome technology and the underdeveloped wisdom of man."[17]

Delgado had a solution to this dangerous situation. He proposed using that awesome technology—developed by humans of underdeveloped wisdom—to modify our own pleasures, sensations, and behavioral responses, and to do it "by direct manipulation of the brain."[18] He pointed out that the technology is here and that important results have already been achieved.

The technology that Delgado had in mind was electricity, and the technique was called Electrical Stimulation of the Brain (ESB): "Because the brain controls the whole body and all mental activities, ESB could possibly become a master control of human behavior by means of man-made plans and instruments."[19] The neurophysiologist demonstrated that small electrical currents flowing through electrodes implanted in the brains of humans and animals produced highly specific behavioral responses, depending upon the specific area of the brain stimulated. Through lengthy experiments, Delgado found the exact parts of the brain in which electrical pulses produced anxiety, fear, rage, pleasure, or euphoria in animal and human subjects.

It was discovered that animals of different species, including rats, cats, and monkeys, voluntarily chose to press a lever that provided electrical stimulation of specific cerebral areas. Animals that initially pressed a lever to get the reward of sugar pellets later pressed at similar or higher rates when electrical stimulation was substituted for food.[20] The animals found so much pleasure in the cerebral stimulation that they totally ignored life-sustaining food in favor of electrical impulses. They became hooked on ESB.

Delgado's research on animals and humans convinced him the time had arrived for humans to achieve a psychocivilized society through technology. He advocated that we declare "'conquering of the human mind' a national goal at parity with conquering of poverty or landing a man on the moon," adding "the project of conquering the human mind could be a central theme for international cooperation and understanding."[21] In the 1960s, the idea of conquering the human mind by means of electricity was appealing and generated considerable interest. An invitation was extended to Delgado to write a volume for the *World Perspectives* Series, whose board of editors included Nobel laureates.

Today, the idea that humans can achieve a psychocivilized society by direct manipulation of the brain through electrical impulses is no longer contemplated. Indeed, scientists are studying the deleterious effects of electricity and electromagnetic waves on humans and other living organisms. The health perils from exposures to electromagnetic fields are wide-ranging—from birth defects to miscarriages to brain damage.[22] The scientific discussion is now focused on the health hazards of electromagnetic fields, not on the achievement of a psychocivilized society through the direct application of electricity to the delicate brain.

So far humans have failed to control any aspect of their behavior on a large-scale basis outside the laboratory confines and theoretical constructs of psychologists. "The daily news tells us again and again that, with all his knowledge and with all his refined ways, modern man remains the wildest animal," lamented Isaac Bashevis Singer, a Nobel laureate in literature.[23]

Starting in the 1940s, B. F. Skinner was sending his utopian messages to the world—that modern humans were now ready to control their own behavior. "This is not time," he said, "to abandon notions of progress, improvement or, indeed, human perfectibility. The simple fact is that man is able, and now as never before, to lift himself by his own bootstraps. In achieving control of the world of which he is a part, he may learn at last to control himself."[24]

The "simple fact" remains, however, that the world of which we are a part is increasing in disorder, uncontrollably. It is true that technologists have put to use a multitude of controlling mechanisms, but these operations are additional entropy-increasing processes. We have not been able to control hurricanes, volcanic eruptions, earthquakes, diseases, insects, floods, droughts, or any general part of the environment; nor have we been able to control human behavior, despite claims that we are ready to do so.

Since World War II, the United States has produced an expanding army of psychoanalysts and behavioral scientists in an attempt to modify, control, and improve human behavior. Yet a survey in the 1940s revealed that the top seven behavioral problems in the public schools were talking out of line, chewing gum, making noise, running in the halls, cutting in line, violating the dress code, and littering. A few decades later the list became drug abuse, alcohol abuse, pregnancy, suicide, rape, robbery, and assault.[25] This is hardly an indication that we are, "indeed," moving in the direction of "human perfectibility."

By removing or ignoring entropy from processes, we project a distorted view of the world, in which we can control anything we wish and make projections into the future in any direction of our choosing. In such a world, the possibilities for shaping the world to our liking become limitless as we bring more and more controlling processes into action.

When scientific discoveries, socioeconomic ideas, or technological breakthroughs are announced, they have hardly any entropy associated with them because they are not yet operational; they are presented as solutions to current and future problems. As we put them to use, we begin to feel their entropies and experience the disorders that they have created. The world does not look the way it was supposed to. It looks more complex, more disorderly, and less controllable.

Is Nuclear Fusion Our Response to the Second Law?

As always, the myth has its charms; but the truth is far more beautiful.[26]

—J. Robert Oppenheimer

In his 1981 book *The Physicists*, Sir C. P. Snow projected a new social hope for humankind. He wrote:

In not much over a generation, physicists have changed our world. That applies to the most elemental of situations, life and death.

Nuclear weapons are an achievement of applied physics. To many people they have brought a new kind of fear. . . .

It won't do any harm, however, to be reminded that applied physics can have an entirely benevolent face. The most dramatic example, as will be seen when this account comes to an end in the year 1980, may be the prospect of abundant energy for ever. If this happens, it will be when nuclear fusion . . . is controlled for peaceful purposes. If this happens, and it is not a certainty, then we shall have a new source of social hope. It is the most exciting promise that applied science has yet suggested—not a firm promise so far, but more than a dream.[27]

Physicists in general are more cautious in raising hope for unlimited energy "for ever." Even the staunchest enthusiast of nuclear energy, Edward Teller, finds the taming of the hydrogen bomb—controlled fusion—a difficult task. He wrote, "Many people who advocate controlled fusion do not understand how difficult it is. They are not the experts, and they promise more than can be delivered."[28] If controlled fusion were a simple undertaking, physicists would have solved the problem by now. They have been working on it for more than 40 years in the United States, Russia, Japan, and Western Europe. The United States alone "has spent upwards of $25 billion pursuing fusion power," according to *U.S. News & World Report*, "with no sure success in sight."[29]

The energy locked in atomic nuclei can be released by two processes—nuclear fission and nuclear fusion. In the first, slow neutrons split the nucleus of a heavy atom—such as Uranium-235—into two lighter fragments, releasing more neutrons and energy. In nuclear fusion, two light nuclei are forced to fuse at ultrahigh temperatures to form a heavier nucleus, releasing neutrons and energy. Energy derived from nuclear fusion is what supports all life on Earth, except that the thermonuclear reactions are happening 93 million miles away, in the Sun.

Why do physicists find controlled fusion so difficult? After all, the Sun has been doing it for a few billion years. Yes, but the Sun's environment is considerably different from our laboratories. To get fusion going requires a huge amount of input energy. For example, in a fusion bomb the temperature needed to start the fusion reaction is attained by using a fission bomb as a trigger. The Sun's trigger is a nuclear reaction that has a very low probability of occurrence. In the Sun, two protons (Hydrogen, H^1) combine to form a mass 2 nucleus (H^2), emitting a positron and a neutrino. In the process, a small amount of mass is lost and that mass is converted to energy, in accordance with Einstein's equation $E = mc^2$.

The "burning" of light hydrogen is a rare event because it goes by the way of the weak interaction, an extremely slow process. The Sun has patience, the kind that can wait for millions of years. In addition, the Sun uses its immense gravitational force to make rare events happen. It has enough matter and pressure to contain a hot, dense center where low probability reactions can proceed. Once heavy hydrogen H^2 (deuterium, D) is formed, the synthesis of heavier nuclides occurs more easily because these nuclear reactions go by the way of the strong interaction, with relatively high probability of occurrence.

We have neither the Sun's mass—gravitational pull—nor its immense patience, so we need reactions that are more efficient. There are two fusion reactions that have a relatively high probability of occurrence. One is the D-D reaction, in which two heavy hydrogen nuclei fuse to form a helium-3 nucleus, emitting a neutron in the process. The other is the D-T reaction, in which deuterium and tritium (T or H^3) fuse to form helium-4, emitting again a neutron.

Because these nuclei have positive electric charges, they repel each other vigorously, resisting the fusing process. Therefore the process requires an inordinate amount of input energy in the form of heat to counteract the inherent electromagnetic repulsive force. At very high temperatures, these gases are completely dissociated into electrically neutral, interpenetrating gases of nuclei and electrons called "plasma."

To make the D-D reaction go requires ignition temperatures approximately 400 million degrees Kelvin, a temperature higher than in the interior of the Sun. The D-T reaction requires lower ignition temperatures, but still close to 40 million K.[30] Because there is no significant natural source of tritium, as there is for deuterium (seawater), tritium must be continuously bred in a nuclear reactor by bombarding the rather rare element lithium with neutrons. Moreover, tritium is a biological hazard because it is radioactive.[31]

Understandably, controlled nuclear fusion has many incredibly hard and complex technical problems. "It may work. It may not," remarks physicist David Rittenhouse Inglis. If controlled fusion does work, "it will have its troubles," he adds, "including radioactive troubles less severe than those of fission reactors."[32] One such headache is stated succinctly by Teller: "Every thermonuclear reaction emits a lot of fourteen-million-volt neutrons. In the long run they tend to destroy almost any material."[33] Whenever the term "clean" source of energy is used in conjunction with fusion energy, it is meant that fusion, compared to fission, "generates fewer kinds of radioactivities," and "less dangerous radioactivities," because the radioactive products cannot be used to make bombs and are shorter-lived than fission by-products.[34]

Fusion also generates much heat. In a nuclear fusion reactor, because of internal pressure, the hot plasma will expand to fill the container in which it is kept. When the hot plasma comes in contact with the walls, it will heat them, requiring the installation of elaborate cooling systems. Eventually we will feel the cumulative effects of the generated heat. Every fusion reactor is equivalent to a mini-Sun transplanted here on Earth.

At present, we speak of fusion as "abundant energy for ever," and "a new source of social hope." However, when controlled fusion becomes operational—if it does—it will also become an abundant source of entropy production. This entropy will not be confined just to the source where the thermonuclear reactions occur, but will manifest itself everywhere the output energy is used. We should ask ourselves, What would this unlimited source of available energy be used for? Probably to produce more cars, more chemicals, and more pesticides. There is also a good probability that it will be used to build more weapons.

Space: The Unlimited Frontier?

It is the last stronghold of the "cowboy," or "frontiers-man," who, after making a garbage dump out of this spaceship, thinks we can solve our population problem by shipping people off to other planets.[35]
— G. Tyler Miller, Jr.

Four decades ago, when humans began space explorations, some futurists envisioned space colonies where we would escape from the earthly problems of economic scarcities, overpopulation, environmental degradation, and nuclear annihilation. The most serious plan was put forth by the theoretical physicist Gerard K. O'Neill of Princeton University. In his 1981 book entitled *2081: A Hopeful View of the Human Future*, O'Neill described his utopian vision:

> The fundamental transformation that space colonies will bring about is from an economics of scarcity—the "zero-sum game" that we are forced to play on Earth—to an economics of abundance. Here on Earth, no nation can enlarge its land area without going to war with another, and every million barrels of oil burned in one country subtracts that much from the reserves all must draw on later. Once we break out from the confines of the planet, we can begin building new lands from the limitless resources of our solar system, and can use freely as much as we need of the sunlight that now streams out, wasted, into the cold darkness beyond the planets.[36]

Space pioneers promise unlimited amounts of natural resources—energy and matter. But so far they have delivered only a few moon rocks for billions of dollars. While they play the "negative-sum game," they vow boundless returns on the investment. It is ironic that at a time when many nations are having difficulty providing home, shelter, transportation, food, and health care to their

people right here on Earth—on the ground floor—some space technologists are promising an abundance of everything up in space.

The fact remains that a spacecraft repair job costs $50 million,
and one space mishap can be disastrous, as happened with the *Chal-
lenger*. And how about the price of a space vacation? "Astronomical!" says the *Los Angeles Times*. If the space shuttle squeezed 50
passengers aboard, NASA would have to charge $10 million per
ticket just to break even, hardly a bargain.[37] Up there in space, the
cost of housing is staggering. The space station, now the International Space Station (ISS), is many years behind schedule and many
times over budget. It was originally envisioned to cost $8 billion and
go into service by 1992. By the time 1992 arrived, however, the
project had gobbled up $10 billion.[38]

The complexity of the venture together with cost overruns have
pushed the estimated ISS cost to $21 billion, on top of the $10 billion
spent on R&D. Here on Earth, even the most elaborate scientific laboratory or the most luxurious mansion costs appreciably less. Maintaining ISS is also costly. According to the Government Accounting
Office, when maintenance costs are included, the ISS project's total
price tag could reach $96 billion for its 10- to 20-year lifespan.[39]

Before we have even had a chance to escape to O'Neill's space
colonies, space has become a dangerous place to live. Hunter killer
satellites are encircling Earth, and some of the "star wars" technologies have been installed up there. Before the desert of space has
been made to bloom into prosperous high-tech space colonies, the
pioneers of this new frontier have transformed it into a gigantic
junkyard where debris and clutter from rockets and satellites
abound. More than 100,000 objects, some of them radioactive, pose
a collision danger to satellites. These artificial meteoroids, primarily
from explosions, hurtle through space at speeds of thousands of
miles per hour. And they are hitting spacecrafts and orbiters. In one
mission alone, the orbiter *Columbia* sustained 106 significant hits,
mainly from human-made particles in orbit.[40]

So as if space exploration were not perilous enough, spacecraft
now circling Earth must be concerned with the rising danger of our
space junk. Because of this situation, the U.S. military is now devoting considerable resources to tracking orbiting space debris. U.S.
spacewatchers use exotic technologies not found in any other
branch of astronomy. The Cavalier Air Station's radar system in
North Dakota, with its 6,144 small antennas, could reportedly detect

a basketball 2,000 miles up. It uses, however, more electricity than a small city.[41]

We have already generated a considerable amount of entropy in space, and we do not yet have hundreds of millions of people circling Earth, nor "200 million people . . . making routine trips into space and back every year," as O'Neill envisioned for the year 2081.[42] The entropy generated by such a large number of space shots would create havoc in the environment. So far we have just been practicing. At any given time, we have had at most a dozen or so people in space, not a few hundred million. Normally, trash and hazardous waste follow in our wake. This time, the debris is way ahead of us, multiplying rapidly and getting ready to greet us in case we really come to stay.

Space pioneers point out the great possibilities that space provides. O'Neill probably shares the sentiment of most space enthusiasts when he says, "We can create in space all the requisites of human existence: air, water, gravity, proper climate, agriculture, and land area. Even more, we can find there the essentials for an advanced civilization: abundant energy and raw materials for industry, and easy pathways for transport and communication."[43]

Here on Earth, we already have ready-made almost everything that O'Neill describes. Yet with all of our money and technological expertise, we have been able to make less than 20 percent of the surface of this planet habitable.[44] Most countries cannot feed their people because they have a poor piece of agricultural real estate. "But let's wave the biggest technological magic wand of all and do something we can't even do on Earth," remarks G. Tyler Miller, Jr. "Let's make all of those unlivable and unappealing planets out there completely habitable."[45]

While it is true that the solar system and the universe at large have unlimited amounts of energy and raw materials, the problem is that for our purposes they are not in readily available or useful form. Humans need to do physical work—expend energy and effort—to get to them and to transform them into life-sustaining goods and products. Thermodynamics tells us that whenever work or energy transformation is involved, entropy is invariably generated in the process. The more processes we invoke, the more entropy we produce. Even with the help of the most sophisticated technologies, it is a much more difficult task to maintain life in space and on other planets because the environments there are not naturally suited for life, requiring additional entropy-producing processes. It is ironic that while modern humanity's agricultural, industrial, and technological processes are gradually transforming planet Earth—a perfectly natural habitat—into an uninhabitable dump, some technologists are preaching that we can transform, through technology, hostile and unlivable environments into flourishing space colonies.

O'Neill and other space enthusiasts are fighting the wrong war. Alluding to the First Law, O'Neill points out that here on Earth humans are forced to play the "zero-sum game": whatever a person or a nation uses is denied to others. We are reminded that we live in an open thermodynamic system, and that our solar system is endowed with abundant energy and matter. However, the First Law is not our most constraining element; the Second Law is a much more formidable hurdle. The principle of conservation of energy merely balances credits and debits, while the principle of entropy prescribes the manner and method of the whole business. Everything we do and everything that goes on in the universe is subject to its stringent constraints.

To open up our thermodynamic system, we need massive technologies, which in turn require energy-matter transformations, generating massive entropy increases in our thermodynamic system. Our environment is already showing signs of exhaustion. The building and maintenance of space colonies would substantially increase our entropy production, and put even more stress on our environment. Besides, wherever we go in this universe the Laws of Thermodynamics will be there.

In 1991, an experiment called Biosphere 2 was conducted in Arizona, at a cost of $200 million, to prove that life can be sustained in hermetic bubbles anywhere within the solar system that electromagnetic radiation would allow. It would demonstrate that we can build ecosystems to our specifications. To many technologists' dismay, the experiment did not support the intended premise—even after a few concessions were made. Two biologists, Joel E. Cohen of Rockefeller University and David Tilman of the University of Minnesota, who analyzed the data as part of an independent team came to the conclusion that "despite its mysteries and hazards, Earth remains the only known home that can sustain life."[46] Which is just what ecologists have been telling us for some time.

Entropy and Growing Global Interdependence

In recent years, growing global interdependence "has had an almost utopian appeal," wrote the *Wall Street Journal* in 1983. The newspaper remarked that this was the dream of the 1960s and early 1970s. Global thinkers maintained that increasing economic interdependence among nations would also bind them together politically. "'Peace through trade' was the catch phrase. A Brazilian-grown chicken in a French-made pot: What a feast the two could make!"[47]

The paper pointed out that global interdependence is now here and growing. It has brought heightened world trade and increased incomes—and living standards—and has propelled some developing countries into industrialization. International investment has bloomed, opening new and seemingly unlimited avenues for the determined entrepreneur.

The increased interdependence also has brought about some problems—disorders—that proponents did not foresee when touting its potential benefits. Nations are far more vulnerable to outside economic disruptions than they were before. Inflation can spread quickly across national borders. Previously safe domestic industries have been hit by world markets, necessitating painful adjustments. Distortions in currency exchange rates can shut out nations' traditional export markets. Economic autonomies of nations have been limited, as drastic actions by one nation quickly translate into hardships to others. "As a result, there are serious questions whether the long-sought-after interdependence hasn't finally reached the point where its costs may be outweighing its benefits," remarked the *Wall Street Journal*.[48]

The desirability of increased interdependence has not always been shared by economists and policymakers. An article in the *Yale Review* called for "a greater measure of national self-sufficiency and economic isolation" among nations. "Ideas, knowledge, science, hospitality, travel—these are the things which should of their nature be international," the author argued. "But let goods be homespun whenever it is reasonably and conveniently possible, and, above all, let finance be primarily national."[49] The author was John Maynard Keynes, in 1933.

"It's still too early to tell how the debate over 'increased inter-dependence' will turn out," concluded the *Wall Street Journal*. "But the concept plainly has far more minuses [disorders] than it seemed to have in the 1960s—and that may require more thought."[50] As it turns out, the Second Law of Thermodynamics gives us an insight into the situation.

Imagine a cube made of a transparent material whose volume is 250 cubic feet, with 250 compartments filled with liquids of different colors. What happens if we make a pinhole on each side of the com-partments? The individual molecules, finding additional degrees of freedom, will start to move around within a larger volume. The en-tropy of the system will increase. When the entropy of a system in-creases, so does our ignorance about the system. Before, we knew that a green molecule was in the green compartment. Now it can be in any compartment. With the passage of time, our ignorance about the system increases as the mixing process goes on. And if the size of the pinhole opening within the compartments should widen, the molecules will find more degrees of freedom to roam around, fur-ther increasing our ignorance—uncertainty—about the system.

The same principle applies to world affairs. Suppose those com-partments were national boundaries. As barriers between nations begin to fall, each constituent (molecule) finds more degrees of free-dom to move around in a larger volume. In our case, the molecules can be anything: people, ideologies, knowledge, religions, raw ma-terials, goods, diseases, chemicals, information (or misinformation), cults, factories, jobs, terrorism, technology, money, food, drugs, or weapons. It is crucial to realize that once physical barriers fall, it be-comes a practical impossibility to "control" the types of things that cross national boundaries.

For example, as international travel increases, diseases spread more easily and rapidly. Not surprisingly, the deadly disease AIDS has, within a short time, become a worldwide problem, because the world has become highly interactive. In this interconnected world, pests, weeds, and dangerous pathogens have also found new ways to move around with ease. Brown tree snakes are hitchhiking from Guam to Hawaii hidden in the wheel wells of a jet. Zebra mussels are being swept up in the ballast water of a supertanker and are finding a new home, and new victims, in the Great Lakes. And the extremely aggressive Asian tiger mosquito, a major carrier of dengue fever, encephalitis, and yellow fever is moving from country to country in containers of used tires.[51]

In 1933, Keynes advocated that knowledge should be international. At the time, humans did not possess the knowledge to build nuclear bombs. Today no one suggests making public the technological know-how to build compact nuclear weapons. And how about the knowledge to manufacture viruses through recombinant DNA: Should that be made international?

We must also recognize that there are vast cultural differences between nations. What some people consider art, others call pornography, and find it offensive. There are also wide differences in social habits. Some nations are high-entropic, oriented toward consumption, while others like to save their money and live a simpler life. Advanced countries with high levels of productivity can produce huge quantities of consumer goods per unit time, and quickly distribute them throughout the world. They can generate intense disturbances all over the world, thus creating additional tensions among nations.

Of course, not all nations are in the same entropic state. When physical barriers fall and the mixing process begins, less-developed nations come in contact with technologically advanced nations, and people with less individual freedom become exposed to those with more freedom. Agricultural nations with simple means of production and communication become cognizant of more intense means of production and faster methods of communication. Pressure mounts on the less-developed nations to "modernize" quickly and become part of the high-entropic community. As more people become indoctrinated into a life-style governed by profligate methods of production and consumption and high levels of obsolescence, the pressure on natural resources heightens, the disorder of the environment magnifies, and the competition—ideological, socioeconomic, and technological—between people and nations intensifies. Life becomes more hectic, more disorderly, more problematic, more uncertain.

The entropies produced by "free trade" and globalization of commerce gradually picked up momentum in the 1980s, especially after the demise of the Soviet Union, and became increasingly disturbing and apparent. Books and articles began to talk about the effects and "minuses" of growing global interdependence.

The discipline of economics accepts the concept of free trade as an axiom not to be challenged. As MIT economist Paul Krugman, author of *Rethinking International Trade*, has pointed out, "If there were an Economist's Creed," it would surely contain the affirmation, "I believe in free trade."[52] Consequently, it is difficult for professors of economics to publish articles and books criticizing the fundamentals behind free trade. Professor John M. Culbertson had to publish his book on this subject himself in 1984.[53] But his ideas have not been ignored, notably by Herman E. Daly and John B. Cobb, Jr. In their book *For the Common Good*, they have elaborated on Culbertson's ideas with cogent challenges of their own to the free trade dogma.

Daly and Cobb remark, "The claim in favor of completely free trade is that it is advantageous not only to both participants but also to the communities as a whole."[54] They show through many examples that this is not the case, as the *Wall Street Journal* pointed out almost two decades earlier. In a later book, *Beyond Growth*, Daly further denounces the economists' free trade dogma: "Economists overwhelming agree that (1) economic growth, as measured by GNP, is a very good thing, and (2) that global economic integration via free trade is unarguable because it contributes to competition, cheaper products, world peace, and especially to growth in GNP. Policies based on these two conceptually immaculate—and interrelated—tenets of economic orthodoxy are reducing the capacity of the earth to support life, thereby literally killing the world."[55]

Free trade with free movement of just about everything among nations has many offensive side effects—from financial to social to ecological. Each book on globalization of commerce emphasizes some aspect of the entropies that free trade produces. In *One World, Ready or Not*, William Greider describes the new world as "dauntingly complex and abstract and impossibly diffuse. . . . Everything seems new and strange. Nothing seems certain."[56] When we increase entropy massively and rapidly, as we are doing through free trade, we also increase the complexity and the uncertainty of the state of the thermodynamic system, as the Second Law implies.

We also cannot ignore the environmental disorders that free trade produces, which most books on the subject point out to some extent. In *The Ecology of Commerce*, businessman and environmentalist Paul Hawken succinctly summarizes the ecological situation this way:

> Quite simply, our business practices are destroying life on earth. Given current corporate practices, not one wildlife reserve, wilderness, or indigenous culture will survive the global market economy. We know that every natural system on the planet is disintegrating. The land, water, air, and sea have been functionally transformed from life-supporting systems into repositories for waste. There is no polite way to say that business is destroying the world.[57]

When factories migrate from industrial to developing nations, a process under way, they bring with them not only capital and jobs but also "environmental dangers." Hilary French of Worldwatch Institute points out, "Hazardous industries, such as battery manufacturers, chemical companies, and toxin-laden computer chip assembly facilities, are becoming increasingly concentrated in countries ill equipped to handle the pollution."[58]

More and more economists are speaking out on the environmental side effects—the externalities—of free trade. In *The Myth of Free Trade*, economist Ravi Batra reminds us that airborne trade pumps millions of tons of jet fuel wastes into the atmosphere; when merchant ships crisscross the globe, they use energy and dump contaminants into the water; when trucks zip through Europe, Asia, and Africa transporting products from one nation to another, they burn gasoline, pollute the air, and contribute to global warming. He concludes: "Thus international trade is a major source of environmental degradation."[59]

Environmental degradation also has a socioeconomic side effect: it is a cause of poverty worldwide. "Poverty is now exemplified by people who search desperately for firewood, find themselves trapped by encroaching deserts, are driven from their soils and forests, or are forced to endure dreadfully unsanitary conditions," writes Wolfgang Sachs in *The Case Against the Global Economy.*[60]

The planet is becoming bitterly divided between rich and poor. Thanks to technological advances, some people and institutions have a tremendous capability to move and manipulate money, information, goods, factories, and jobs around the world at dizzying speeds, thus creating entropy at unprecedented levels. In *Divided*

Planet, environmental activist Tom Athanasiou points out, "From
the perspective of the new world disorder, with economic, political,
and ecological chaos all competing for our scant time and attention,
. . . care for the weak and vulnerable, environmental protection, or
even democracy, the schemes of the post-World War II geopoliti-
cians do not seem to have worn well at all."[61] What many are asking
is, Who is going to care for the weak and vulnerable, and the environment that we all depend on for our survival?

While some enthusiasts of the free trade doctrine find globalization a near Utopia for humanity, others point out the entropies
generated by its processes.[62] As John Gray, professor of European
thought at the London School of Economics, tells us in *False Dawn:
The Delusions of Global Capitalism*, already the "Utopia of the global
free market" has resulted in more than "a hundred million peasants
becoming migrant labourers in China, the exclusion from work and
participation in society of tens of millions in the advanced societies, a condition of near-anarchy and rule by organized crime in
parts of the post-communist world, and further devastation of the
environment."[63]

The socioeconomic disorders generated by increased global free
trade have become so pronounced that even George Soros, a long-
time capitalist and practitioner of international finance, is writing
about them. Recent events, such as the financial collapse of Asia,
the Russian meltdown, and the financial crisis in Latin America,
made Soros even more aware that "the global capitalist system was
unsound and unsustainable."[64] In *The Crisis of Global Capitalism*, he
remarks that classical economists were inspired by Newtonian
physics and its laws. Their goal was to use the laws of mechanics to
explain and predict economic behavior.[65] As Soros points out, this
has not happened:

> The rethinking must start with the recognition that financial mar-
> kets are inherently unstable. The global capitalist system is based
> on the belief that financial markets, left to their own devices, tend
> towards equilibrium. They are supposed to move like a pendulum:
> they may be dislocated by external forces, so-called exogenous
> shocks, but they will seek to return to the equilibrium position.
> This belief is false. Financial markets are given to excesses and if a
> boom/bust sequence progresses beyond a certain point it will

never revert to where it came from. Instead of acting like a pendulum financial markets have recently acted more like a wrecking ball, knocking over one economy after another.[66]

Soros is expressing what thermodynamics has been telling us through its Second Law for nearly a century and a half—that we live in an irreversible world, not a pendulum-like world. In the early 1970s, economist Nicholas Georgescu-Roegen began urging his colleagues to pay attention to the Law of Entropy. Herman Daly has reminded economists of his teacher's work.[67] In *Living Within Limits: Ecology, Economics, and Population Taboos,* Garrett Hardin reiterates in strong words that economists can no longer evade the Laws of Thermodynamics.[68] Had economists embodied the principles of thermodynamics in their theories, they would not have made such a gross error in thinking that financial markets, or economic activities in general, behave like a reversible mechanical pendulum.

Chapter 10

The World Through the Eyes of Thermodynamics

Progress is the mother of problems.[1]
—Gilbert K. Chesterton

The Concept of "Doing More with Less"

Nothing recedes like progress.[2]
—e. e. cummings

Technologists espouse the idea that humans are doing more and more with less and less. They maintain we are achieving ever-increasing efficiencies in the use of matter and energy. Buckminster Fuller coined the word ephemerilization for this "progressive do-ing-more-with-less" concept.[3] Technologists cite situations where energy-matter efficiencies have been achieved through technological advances. For example, when Charles Lindbergh crossed the Atlantic in 1927, his airplane had an empty weight of about 2,150 pounds and cruised at an average speed of 100 miles an hour. Today an airplane of comparable dimensions weighs a few hundred pounds less and cruises at faster speeds. "Although today's airplane flies faster, it uses less fuel per mile," writes scientist J. Peter Vajk. "The differences between the two aircraft are attributable to improvements and advances in materials, science, and engineering techniques."[4]

There is no question that technological inventions lead to improvements in efficiency. These come about through massive infusions of energy and effort in R&D. The research, development, and production of passenger jets, due to increasing complexity, have become so prohibitively difficult and expensive that only one commercial aircraft manufacturer remains in the United States—Boeing. In Western Europe, not a single nation has the resources to sustain R&D efforts in commercial aviation. Consequently, governments and businesses have formed a consortium—Airbus—to manufacture wide-bodied jets. Aircraft manufacturers might well wish that the concept of doing more with less could apply to R&D efforts.

As discussed in Chapter 2, French military engineer Sadi Carnot's energy efficiency principle sets a theoretical limit on the maximum efficiency of any engine—whether it is an internal combustion engine, a Diesel engine, or a steam engine. In the case of the automobile, we have recently achieved noticeable improvements in energy efficiency. Today's cars get more miles per gallon than their predecessors, but they are also much smaller and much lighter; they are not the same cars we used to drive. Carmakers have gradually replaced metal and wooden parts with plastic, which has brought about some fuel economies. Furthermore, to gain greater fuel efficiency, carmakers have added under the hood a host of computerized cybernetic gadgets, which in turn have increased the complexity of the car, the headaches of car maintenance, and the overall entropy of our thermodynamic system.

While it is true that airplanes and automobiles have become more energy-efficient over time, they are not, thermodynamically speaking, the most efficient means of transportation. Railroads, buses, and mass transit in general have higher energy efficiencies per unit of passenger traffic, while bicycles consume practically no energy. We use cars and airplanes instead of bicycles, buses, and railroads for convenience and speed of travel, but we are paying a price to the Laws of Thermodynamics. A relatively small and "efficient" car, such as a Ford Escort, uses 45 percent of the available fuel energy to cool its engine, and dissipates another 38 percent in the exhaust system, normally known as pollution. The remaining 17 percent of the energy is used to propel the car and its passenger and perform other activities.[5] Overall, the passenger taps only an insignificant portion of the energy, while the greatest majority of the fuel goes into waste and heat. The automobile is an excellent illustration of Carnot's energy efficiency principle in action.

With the advent of sport utility vehicles and minivans, the trend toward higher miles-per-gallon vehicles has reversed in the United States. While the average fuel economy of new U.S. automobiles nearly doubled between 1974 and 1988, the average fleet efficiency has gradually declined since then. These bulky gas-guzzlers have now become so popular that they account for 45 percent of all new car sales in the United States.[6]

"Similarly, soaring sales of high-wattage halogen 'torchiere' floor lamps are negating efficiency improvements in the lighting sector," comments *World Watch* magazine. World sales of energy-efficient compact fluorescent lamps (CFLs) increased eightfold between 1988 and 1997. CFLs reduce pollution from electricity generation and save money. Since the early 1990s, however, halogen floor lamps have become popular in the United States. Consequently, the 40 million halogen lamps in use in the United States are now consuming more electricity than the 280 million CFLs are saving, notes *World Watch*.[7]

Technological advances provide us with many new conveniences, but also create many opportunities for waste. As energy-efficiency advocate Amory Lovins points out, "the United States runs the equivalent of five nuclear reactors solely to power devices that are turned off, such as color televisions in standby mode."[8]

Technology enthusiasts use the ephemerilization concept to promote the speeding up of technological "progress." In the words of economist and *Wealth and Poverty* author George Gilder:

> Technological progress throughout history has entailed the replacement of heavy and potentially damaging machinery with more efficient and less environmentally destructive means of production. The obvious development is from the steam engine to the silicon chip. Current microprocessors the size of a fly have more computing power than the early computers that would fill up a gymnasium with tubes and wires. In general, economic progress has tended to result in smaller, lighter, more efficient, less pollutant equipment. To hold back economic progress in the name of reducing environmental damage is entirely counterproductive.[9]

We should not forget, however, that before computers came along we had slide rules. They required no electricity to run and lasted practically a lifetime.

In thermodynamics, what matters is the total entropy increase in the thermodynamic system. Small and light objects can generate a lot of disorders. Today's nuclear weapons are not very large, yet we all know of their destructive power. As professor Witold Rybczynski has pointed out, "Aerosol cans are very small, and yet they've created a very large problem. Those who argue that unintended consequences can be avoided through small technology are promoting a myth."[10] In the case of the silicon chip, manufacture requires gases, chemicals, and solvents, which in turn become part of our environment, thus increasing the entropy of our planet. The production of one 15-centimeter silicon wafer—from which a few dozen chips are cut—demands elements that are many times its size and weight, such as 9 kilograms of liquid chemicals, 6 cubic meters of gases, and 8,610 liters of water.[11] As Max Planck reminded us repeatedly, we need to account for the total thermodynamic situation, not part of it.[12] The discipline of thermodynamics keeps us honest.

When computers were bulky and expensive, they were only a few of them, and they were mainly used to solve complicated equations of physics and engineering. Today, thanks to technical innovations, we have millions upon millions of microchips and computers doing every conceivable job, from the most complex to the most trivial. In addition, their numbers are multiplying fast, generating not only environmental disorders but also informational entropy. Ironically, "it is precisely the economic gains from improved technical efficiency that increase the rate of resource throughput," write regional and community planners Mathis Wackernagel and William Rees. "Far from conserving natural capital or decreasing Ecological Footprints, this leads to competitively accelerated increases in consumption."[13]

The consumption of materials keeps going up because the number of cars, telephones, computers, and air conditioners increases faster than the material reductions gained by efficiencies in materials usage. For example, while the weight of mobile phones was reduced 10-fold between 1991 and 1996, the number of subscribers to cellular phone service jumped more than 18-fold. Moreover, these mobile phones were typically additions to a household's telephone inventory, not merely a replacement of old equipment.[14]

Computers were supposed to reduce paper consumption. In computer circles, the "paperless office" became a paradigm. However, as Edward Tenner tells in *Why Things Bite Back*, this dream did not materialize. Tenner saw something curious happening where he worked, an academic publishing house. As personal computers appeared on one desk after another, the paper recycling bins always looked full. Even after the office was networked and e-mail had replaced hard-copy communications, "the paper deluge continued." He also noticed that office suppliers had taken advantage of the situation, and were displaying five-thousand-sheet cases of paper for photocopiers, printers, and fax machines. This phenomenon was the motivating force for Tenner's book.[15]

In the United States where computers are practically everywhere, per capita consumption of paper exceeds 725 pounds a year, nearly seven times the global average.[16] Since the advent of the personal computer, office paper purchases have surged from 4.4 million tons in 1980 to 7.3 million tons in 1998.[17] Not surprisingly, "some analysts consider the rise in electronic communications to have been 'a great blessing to the paper industry,'" notes Worldwatch Institute.[18]

If the doing-more-with-less concept, also known as "dematerialization," had an effect on overall materials usage, the United States—a major player in technological innovations—would have been a beneficiary; per capita use of materials would have continually shrunk. The data show otherwise: from 1900 to 1995, while the U.S. population increased less than 4-fold, the U.S. materials consumption grew 18-fold.[19] Globally, the situation is not much different on the energy front, the other side of the energy-matter equation: from 1900 to 1997, while world population increased less than 4-fold, world energy consumption grew more than 10-fold—with a quadrupling since 1950.[20] Thermodynamically speaking, we are doing more with more materials, with more energy, and with more people; that is, we are producing more entropy than ever before.

Technological advances have not necessarily brought about smaller, less polluting equipment. Nuclear power plants, a result of modern technology, are not small, and they generate long-lived, highly radioactive materials in addition to heat: The nuclear debris contains uranium-235 and plutonium, which, ounce for ounce, is the most toxic material on Earth.[21] Aircraft carriers are gigantic ships that consume huge quantities of energy on a daily basis. In the case of the chemicals, which we use for technology and other purposes, their size and energy efficiency are not important parameters. Because every chemical process increases the disorder of our thermodynamic system, the volume of output and toxicity of chemicals are more significant variables. Toxic-waste chemicals are quickly filling our landfills. And in the case of such chemicals as fertilizers and pesticides, we use more of them with the passage of time because they lose their effectiveness. The idea of doing more with less surely does not apply to these chemicals, while the Laws of Thermodynamics invariably do.

"Indeed, despite improvements in efficiency in recent decades," writes environmental critic Joshua Karliner, "the chemical industry still produces more waste than product." For example, "chemical production generates two barrels of waste for every barrel of product. . . . For pesticides the ratio is three units of waste to one unit of product. For synthetic dyes it is eight to one."[22] These wastes are generated during production. When we put chemicals, pesticides, and fertilizers to use, we have additional entropies to deal with. The Second Law collects its dues both during production and during consumption.

There is a school of thought closely related to the doing-more-with-less concept, which contends that through the power of the mind, humankind overwhelms the laws of matter and energy. Consequently, the laws of physics and chemistry become inconsequential, especially in economic processes. In the words of George Gilder, a campaigner for this ideology, "because economies are governed by thoughts, they reflect not the laws of matter but the laws of mind."[23]

While it is true that human thought processes can wander in any imaginable direction when unconstrained, once they are put into action, whether in economics, technology, or other fields of human endeavor, they collide instantaneously with the Laws of Nature. The laws of mind, whatever they may be, do not operate above the Laws of Nature, as is so fondly believed in some quarters. In 1989, Gilder explained the essence of this belief in this fashion:

> Gone is the view of a thermodynamic world economy, dominated by "natural resources" being turned to entropy and waste by human extraction and use. Once seen as a physical system tending toward exhaustion and decline, the world economy has clearly emerged as an intellectual system driven by knowledge. The key fact of knowledge is that it is anti-entropic: it accumulates and compounds as it is used. Technological and scientific enterprise, so it turns out in the age of the microcosm, generates gains in new learning and ideas that dwarf the loss of resources and dissipation of energy. The efforts that ended in writing E equals MC squared or in contriving the formula for a new photovoltaic cell or in inventing a design for a silicon compiler . . . are not usefully analyzed in the images of entropy.[24]

Speaking of the Second Law, Gilder says: "Capitalism, with its imperative of growth, can be depicted as violating the very law of nature."[25]

This is not the case. As far as physics is concerned, we are still living in a world governed by the Laws of Thermodynamics. Capitalism, or any other system, cannot grow without increases in entropy. Moreover, knowledge, like everything else, is not "anti-entropic." The word anti-entropic is not in the vocabulary of physics. Indeed, human knowledge has not produced a perpetual motion machine of the second kind, an act that would have demonstrated beyond a shadow of doubt that knowledge is anti-entropic. Basic science, however, has provided us with the knowledge that all processes increase entropy and that it is impossible to create anti-entropic processes by any means whatsoever. This scientific message has remained constant and unshaken ever since Lord Kelvin and Rudolf Clausius formulated it about a century and a half ago.

Knowledge is the very source of technological and product developments, which generate entropy. We could choose to talk only about the equation $E = mc^2$ and not mention the radioactive waste products that nuclear reactions produce while transforming mass to energy. We could choose to write only about the amazing microchip, never mentioning the gases and chemicals required to produce it. We could choose to shower people with technological breakthroughs, never telling anybody about their side effects. But science, which is based on discovering and upholding the *truth* and the *whole* truth, would be derelict if it attempted to boast of its great discoveries and never told the public about the consequences of their applications.

We have reached a point in our thermodynamic evolution where we can no longer ignore the entropies generated by our intellectual activities and economic processes. The accumulated entropies have become too obvious—even to untrained observers. We can no longer deny the existence of toxic chemicals, radioactivity, electropollution, pesticides, and smog in our thermodynamic system. Nor can we ignore the disorder in our knowledge and information base.

Change and Technological "Progress"
Re-examined

*Instead of conveying a teleological quality, the word "prog-
ress" now means just moving on, even though the forward
motion is on a road that leads to disaster or despair.*[26]
—René Dubos

*We have taught ourselves to create and combine the most
powerful of technologies. We have not taken pains to learn
about their consequences. Today these consequences threaten
to destroy us. We must learn, and learn fast.*[27]
—Alvin Toffler

The fast pace of change is becoming increasingly apparent. In 1970,
in his celebrated book *Future Shock*, Alvin Toffler recognized this
fundamental problem of modern life. He pointed out that today we
are being overwhelmed by the high velocity of changes taking
place in our environment. "Change is life itself," Toffler wrote. "But
change rampant, change unguided and unrestrained, accelerated
change overwhelming not only man's physical defenses but his
decisional processes—such change is the enemy of life."[28] The
noted futurist summarized his thoughts with these words: "Future
shock—the disease of change—can be prevented. But it will take
drastic social, even political action. No matter how individuals try
to pace their lives, no matter what psychic crutches we offer them,
no matter how we alter education, the society as a whole will still
be caught on a runaway treadmill until we capture control of the
accelerative thrust itself."[29]

How can we take charge of this accelerative change and prevent
future shock? Toffler did not offer an overall solution, acknowledging that the basic thrust of the book was diagnosis.[30] He did identify
technological advance as a "critical node in the network of causes,"

and perhaps the node that activates the whole network: "One powerful strategy in the battle to prevent mass future shock, therefore, involves the conscious regulation of technological advance."[31]

A decade later, in *The Third Wave*, Toffler wrote: "While *Future Shock* called for certain changes to be made, it emphasized the personal and social costs of change. *The Third Wave*, while taking note of the difficulties of adaptation, emphasizes the equally important costs of not changing certain things rapidly enough."[32] In Toffler's view, humankind has undergone two major waves of change. The First Wave—the agricultural revolution—lasted a few thousand years; the Second Wave—the industrial revolution—lasted about 300 years; the Third Wave, which is even more accelerative, may be completed in a just few decades.[33]

What is the Third Wave all about? "It is, at one and the same time, highly technological and anti-industrial," explains Toffler.[34] In fact, the foundation of the Third Wave is high technology itself, propelled by computers and telecommunication systems. Toffler describes the power of the computer in this fashion:

> The computer not only helps us organize or synthesize "blips" into coherent models of reality, it also stretches the far limits of the possible. No library or file cabinet could think, let alone think in an unorthodox fashion. The computer, by contrast, can be asked by us to "think the unthinkable" and the previously unthought. It makes possible a flood of new theories, ideas, ideologies, artistic insights, technical advances, economic and political innovations that were, in the most literal sense, unthinkable and unimaginable before now. In this way, it accelerates historic change and fuels the thrust toward Third Wave social diversity.[35]

What was the original problem called future shock? It was the premature arrival of the future, thrust by rampant, unrestrained, and accelerated changes, which Toffler called the "enemy of life." And what is the amazing computer doing? As Toffler admits, it is fueling that acceleration.

The question remains: Is the computer the enemy of life or a tool that helps us organize? Is it the disease itself, or the cure? To answer these questions objectively, it is essential that we analyze the phenomenon of future shock in terms of the Laws of Nature. In life, we experience the effects of natural laws; we feel time dissipating because of entropy increases in our thermodynamic system. The more massive the entropy increases around us and within us, the more we experience time flowing faster and the future arriving earlier.

We are not suffering from future shock by accident, but because we have become high-entropic creatures. We are introducing high technology at a fast pace, with the overall view that we are "progressing" through technological advances. "We have trained our social reflexes for technological 'advances,' however trivial their goals and deleterious their long-range effects," observed ecologist René Dubos.[36] The idea of technological progress is entrenched in the deepest thought-processes of nearly everyone alive today.

In *Future Shock*, for example, even after identifying technological advance as a major contributor to future shock, Toffler said "we clearly need not less but more technology."[37] Of course, people have been doing all along what Toffler is advocating—introducing more and more technology. The stubborn fact remains, however, that we have been experiencing problems with each new technology, as stated succinctly by Toffler himself: "Our technological powers increase, but the side effects and potential hazards also escalate."[38] That is the thermodynamic catch. Future shock is not only about the speedy arrival of tomorrow, but also about the environmental, social, informational, political, and economic disorders all around us. Future shock is about high entropy in our thermodynamic system.

Toffler has not been the lone voice narrating the side effects of technology. Technologists themselves have become the most articulate spokespersons of this theme. In 1964, at his inaugural address as President of the Massachusetts Institute of Technology, Julius Stratton voiced the following concerns to the faculty and students:

> Diseases of the system are emerging in increasing number; and we must be courageous in recognizing that they are themselves the by-products of our highly technological environment.
>
> Consider the transformation of our cities—the physical and social degradation of large areas—the loss of serenity and beauty. We have never before produced so many cars or such fast airplanes; yet transportation in the United States is rapidly approaching a point of crisis. The shift to automation in industry is accelerating and will have profound effects upon the character of our labor force, upon its training, and upon its security. We are polluting our

air and water. The pesticides which we are employing on a mounting scale are a boon to agriculture and a threat to the remainder of our natural resources.[39]

Expressions of concern by luminaries about the side effects of technology are now commonplace. Whenever this happens, the discourse has a certain soothing effect on society. It gives the impression that the technological establishment has finally recognized the existence of the disease, and a cure will follow in due course. But problems persist, and the diseases of the system are multiplying and spreading even faster than before. The reason is that our attitude toward the problem has not changed. While we recite the side effects and hazards associated with modern technology, our posture on the method of approaching the problem remains rigid. We advocate solving the disorders generated by technology through technological advances. Consequently, society at large is being submerged in technological complexity and entropy.

Scientific endeavors begin by asking why a phenomenon occurs the way it does, and then by coming up with an explanation—usually with a Natural Law that governs the set of observed phenomena. Technology, on the other hand, asks such questions as what to build, how to do it, and when does it need to be done. Technology is not involved in explaining why diseases of the system are emerging in increasing number, or why we are experiencing future shock, or why the side effects of technology are escalating as our technological powers increase.

Magazines report the side effects of technologies. But when the solution is put forward, it usually involves more technology. For example, an article on technology in the *Wall Street Journal* starts with a problem: "The information glut is one of the curses of the electronic age. News streams constantly from television, radio, news wires and electronic-information banks. But the customers have to plow through piles of dross to find the nuggets they want." And the solution to this information entropy follows immediately: "Electronics is coming to the rescue. Companies are developing systems that will let viewers, listeners and readers select precisely what information they will get."[40]

This is a classic example of the very technology that has created the disorder being flipped to become our savior. In *The Technological Society*, French sociologist Jacques Ellul points out, "History shows that every technical application from its beginnings presents certain unforeseeable secondary effects."[41] And when the secondary effects—disorders—become noticeable and unbearable, we often introduce, by necessity, another technology to alleviate the problem.

In recent history, advances in computers and communication systems have accelerated the growth in information entropy. Solutions in information systems abound, yet computer-related problems are piling up, prompting *Informationweek* to ask: "If Everyone Has 'Solutions,' Why Are There So Many Problems?" The editor notes that solutions "are offered not only *by* computer companies, but *to* them as well."[42] Information entropy knows no boundaries.

During the last quarter-century, while technological solutions and advances in computers were introduced at a dizzying rate, problems in data processing not only multiplied but also increased in severity.[43] Every technological field has been experiencing more or less the same phenomenon. That is why our problems have been increasing steadily, as well as becoming more complex. In thermodynamics, all processes contribute somehow and somewhere to the disorder of the thermodynamic system—even those we conveniently classify as problem-solving technologies.

In technological societies, the rapid pace of technological change thrust by computers has brought about considerable stress and mental anxiety. We now have some clinical psychologists who specialize in mental disorders emanating from technology-related stress. Craig Brod, a Silicon Valley psychotherapist, coined the term "techno-stress" for a contemporary ailment caused by the pressures of living in a high-tech world.[44] Its symptoms include "tension, paranoia, overstimulation, anticipatory disaffiliation, psychosomatic headaches, fatigue, sagging libido, psychic numbing, low self-esteem and high anxiety."[45]

Many global thinkers believe that the twenty-first century will inevitably be a period of unprecedented technological progress, which, in conjunction with globalization, will spur a powerful and sustained economic growth. They maintain that we have barely scratched the surface in such technologies as genetic engineering and microphysics. The next wave of technological progress will be even more rapid, more robust, and more pervasive. This in turn will induce faster growth rates, which will make it much easier to address some of our vexing social and environmental problems.

In a special double issue on the economy of the twenty-first century, *Business Week* echoed this belief. The theme was "You Ain't Seen Nothing Yet," predicting even faster rates of technological progress, accompanied with rapid and massive distribution of new products throughout the global economy.[46] Thermodynamic translation: "You Ain't Seen Entropy Yet." If such a prediction materializes, whereupon entropy production truly explodes, today's complex, uncertain, and disorderly world will seem by comparison like the simplicity, stability, and serenity of the 1950s.

It is becoming increasingly obvious to some technologists that modern problems will not be solved through technological breakthroughs. "One thing seems clear," says Paul B. MacCready, Jr., the father of human-powered flight, "Getting through this stressful time is not going to depend on any technological advance—an improved antimissile system, better computers and communications, free energy or a cure for cancer. It will depend on using our brains to get along better with one another and our globe."[47] (If energy were indeed free, entropy production would explode.)

MacCready made his feelings clear: "Unless we find new approaches, I am very pessimistic about the future of civilization." He pointed out that we are in the process of wiping out a good portion of our natural resources, "most of the rain forests, and a huge number of plant and animal species. That doesn't include the threats of nuclear weaponry, ecological pollution and genetic tampering," adding: "our brains are the problem, but they also represent the only possible solution."[48]

The problem is not the biochemical composition of the brain. Through millions of years of gradual evolutionary changes, Nature has done a marvelous job in producing our brains. But now the brain is in a state of high entropy: inconsistent, incoherent, and incorrect messages are bombarding it at a mind-numbing pace. Moreover, thanks to technological advances, the disordered messages—arriving chaotically from every direction—are hitting the brain at an earlier and earlier age, generating confusion and entropy.

Although there may be certain things that we must change rapidly, this does not include all upcoming technologies—no matter how seductive they may appear to some of us. As former State Department official Lindsey Grant remarks, "Humankind right now needs a coherent vision of where we are headed and a way of considering the consequences of technological change, much more than it needs new technology."[49] It does not hurt to remind ourselves that the dinosaurs lasted tens of millions of years without any technology. What we need to change rather quickly is our attitudes, our way of thinking, our life-styles, and our view of how the world works. "Things do change," mused Isaac Asimov. "The only question is that since things are deteriorating so quickly, will society and man's habits change quickly enough?"[50]

Recognizing Low- and High-Entropic Actions and Life-styles

When President Calvin Coolidge left office he told reporters, "Perhaps one of the most important accomplishments of my administration has been minding my own business."[51] In the eyes of many critics, "Silent Cal" was a do-nothing President, and for that reason history has not been kind to him. He is regarded as one of the most "inexplicably maligned Presidents."[52] If we judge him under the Laws of Thermodynamics, however, we find him to be a low-entropic person who generated hardly any disorder, and his record shows it. Viewed "without ideological blinders," states Coolidge scholar Thomas Silver, "Coolidge's record looks pretty good: he balanced the federal budget, cut the national debt almost in half and kept inflation at only 0.4 percent—all the while presiding over a 17.5 percent growth in the gross national product."[53]

Of course, when Coolidge presided—in the 1920s—the entropy of the world, including the United States, was much lower. Nevertheless, he was a minimalist with a profound belief in economy of word, deed, and government, which he held to throughout his public life. He once quipped that it was in the national interest that he slept so much (an average of 11 hours a day), since he couldn't ·create any disorder when he was asleep. His low-entropy philosophy of life is indicated by his statement that "there is no dignity quite so impressive, and no independence quite so important, as living within your means."[54] Many presidents have not followed Coolidge's philosophy, and the federal government has amassed a huge national debt—unpaid entropy.

In everyday life, we come across numerous situations in diverse areas—from politics and health care to agriculture and economics—where the discipline of thermodynamics can guide us on a particular course of action. From the Laws of Thermodynamics we can deduce a practical philosophy of life. In health care, for example, suppose a person is diagnosed as having high blood pressure. A high-entropic approach would be to continue living the same lifestyle and to control the high blood pressure through drugs. But the Second Law tells us that all processes, including medicine, increase entropy—that is, produce some side effects. A probability exists that another medical condition may develop, requiring yet another medical intervention to correct the new medical disorder.

A low-entropic approach would be to first try eliminating the problem naturally through such things as reducing salt intake, proper diet, and loss of extra weight, as well as appropriate physical exercise. Under this approach, there is no dependence on drugs and no risk of their side effects Thermodynamics advises us to take medicine only when necessary.

It turns out that good health and longevity can be achieved by simple low-entropic health practices, referred to as "clean living." A study conducted by the California Health Department isolated seven "golden rules" of behavior that, if followed, promote a longer life expectancy: eat a regular breakfast every day; don't smoke; eat three meals a day, regularly; get about eight hours of sleep each night; keep a normal weight; drink no more than one or two alcoholic beverages a day; and exercise regularly, particularly in active sports. As Dr. Lester Breslow, an author of the study, remarked, "The daily habits of people have a great deal more to do with what makes them sick and when they die than all the influences of medicine."[55]

A high-entropic approach to health care involves the development of all kinds of biomedical intervention for every conceivable health disorder, followed by an aggressive campaign for their promotion. The Second Law warns us that even the controlling mechanisms (processes) increase entropy; we have to be aware of this fact. Science tells us about the Laws of Nature and how Nature works, leaving the choice to us—individuals and society—as to how we want to conduct our lives.

Conducting scientific research on how to prevent medical disorders and promote healthy habits is a low-entropic activity. The old adage "an ounce of prevention is worth a pound of cure" has a low-entropic ring to it—with economic importance. For example, the cost of prenatal care for a healthy pregnant woman, for nine months, is about $600, whereas the cost of medical care for an extremely premature baby is about $2,500 a day; a measles shot costs $8, whereas hospitalization for a child with measles costs $5,000.[56]

We can develop low-entropic eating habits by consuming foods that have undergone a minimum amount of transformations. For example, if we eat more fresh foods and less canned and frozen foods, we conserve energy and produce less entropy. If we eat more grains, vegetables, and fruits, and less beef, pork, and lamb, we contribute less to the overall entropy production. Since cattle, hogs, and sheep feed on grains and plants, they are higher-entropic food. It takes about 22 calories of plant energy in the form of feed to produce a calorie of food energy as beef.[57] Consequently, our dues to the Second Law are much higher when we eat meat, which is reflected in its price. "The second law makes it clear why a pound of steak costs so much more than a pound of corn or wheat," writes chemist G. Tyler Miller, Jr., adding that a diet rich in meats "represents a massive waste . . . not to mention the massive pollution that modern agriculture directly and indirectly imposes on the environment."[58] In the United States, livestock produce about 1.37 billion tons of waste annually.[59]

Regarding eating habits, a low-entropic message comes from simplicity advocate Janet Luhrs. In *The Simple Living Guide: A Sourcebook for Less Stressful, More Joyful Living*, she writes: "*Eat the food in its closest state to the way it came out of the ground.* No fuss, no muss." She reminds us that food loses nutrients with each process it undergoes. By the time most food has reached the grocery store in a beautifully designed, colorful box or can, it has almost no vitamins left. Moreover, we are paying people to process the food, add chemicals to it, design the box, and advertise the product to make us buy it.[60]

In agriculture, we can adopt environmentally less damaging—lower-entropic—means of food production, like organic farming. Studies conducted by the U.S. Department of Agriculture in 1980 and the National Research Council in 1989 revealed that organic agriculture, which avoids or largely excludes the use of pesticides and synthetically compounded fertilizers, is a viable alternative to high-intensity agriculture. They found that organic farms on average are somewhat more labor-intensive but use less energy than conventional farms, and that "well-managed farms can get the same or better crop and livestock yields with natural techniques, compared with farms using heavy doses of synthetic fertilizers and pesticides."[61]

Professor of Biological Sciences Mark L. Winston tells us that we need to change our philosophy in the war against pests and adopt a lower-entropic approach: We should try to "manage rather than control, reduce instead of eradicate, tolerate rather than panic."[62] His philosophy stems from the observation that pesticides are poisonous and are damaging our environment and posing health hazards. Moreover, many of the pests are near impossible to control, let alone eradicate. Cockroaches provide an excellent case in point. Contemporary roaches are primitive insects, relatively unchanged from the ones that walked Earth 350 million years ago, long before the dinosaurs, "and there is every reason to believe that cockroaches will continue to exist on earth in more or less their present form long after humans have become extinct."[63]

Winston points out that "their biology is elegant in its simplicity and is ideally adapted for life as a human pest. Cockroaches are flat, long-legged, and greasy; they flee from light, prefer moist environments, protect their young, and eat almost anything."[64] Their scavenging life-style is perfect to our throwaway society. We, on the other hand, are complex biological entities demanding high levels of maintenance—both physically and mentally. Despite our advanced evolutionary state, we are not suited to win the war against biological creatures like cockroaches. Consequently, we should heed Winston's advice and change our ethics toward pests. Besides, as Winston reminds us, "our war against pests has exacted a heavy toll."[65]

To reduce environmental disorders and waste, we need to reduce energy transformations and processes. Recycling or reprocessing materials helps, but we need to do much more. "Though reprocessing is good, prevention of premature obsolescence is better," notes Edward Teller. "Washing machines should not have to be scrapped because one small part cannot be replaced. Today there is not much choice. Spare parts are not generally available for old equipment, and repair has a hard time competing with replacement."[66]

Teller is pointing out that low-entropic life-styles are having a tough time competing with high-entropic ones, and that obsolescence is deeply rooted in the American way of life. Through globalization of commerce, this life-style is now spreading to other countries. In 1960, in *The Waste Makers*, journalist and social critic Vance Packard described how the American economy thrives on obsolescence, a process that has only picked up momentum since then. Packard called the process "progress through planned obsolescence."[67] The high rate of obsolescence we are experiencing today applies not only to equipment and machinery but to practically everything—from clothing and information to technology and ideas. As Marcus Cunliffe, an authority in American history, has remarked: "American society has a quite extraordinary, excessive fascination with novelty. The nation keeps chucking out old ideas and puts a premium on having bright, new ones. . . . New ideas too often are exaggerated reactions to other ideas—with each, in turn, discarded as fashion changes. That is not a good way to operate."[68]

Ideas, when put to action, create new processes that, in turn, render obsolete the existing processes, technologies, information, clothing, and machinery, thus generating entropy and waste. Moreover, the faster the rate of obsolescence and change, the higher the production rate of entropy. The result manifests itself as disorders in the environment and the socioeconomic system.

As described in Chapter 5, the discipline of thermodynamics also tells us that machines, while generating entropy, dissipate time. Thus if we desire to have more time to ourselves, we must reduce the number of machines in our possession, including the ones we conveniently label as "timesaving" devices. The Amish people have plenty of leisure time precisely because they have voluntarily stayed away from a great number of timesaving modern conveniences. Their way of life exemplifies a low-entropic life-style amidst today's high-entropic world. "Anyone stepping into Amish society suddenly feels time expand and relax," notes sociologist Donald Kraybill, who spent time among the Pennsylvania Amish.[69]

"The great irony here is that in Amish society, with fewer labor-saving devices and other technological shortcuts, there is less 'rushing around.' The perception of rushing seems to increase directly with the number of 'time-saving' devices," observes Kraybill. "Although much time is 'saved' in modern life, for some reason there is less of it and rushing increases. The perception of rushing increases as the number, complexity, and mobility of social relations soar. Thus the simplicity, overlap, and closeness of Amish life slow the pace of things and eliminate the need for time-management seminars [which further dissipate time]."[70] As far as the discipline of thermodynamics is concerned, there is no irony to the fact that the Amish have more leisure time while having fewer labor-saving devices. By limiting the use of energy-consuming and entropy-generating technologies, the Amish have reduced their dues to the Laws of Thermodynamics. Their basic values—"work is more satisfying than consumption" and "newer, bigger, and faster are not better"—promote a low-entropic way of life.[71]

Even more intriguing than the Amish is the case of some tribal and indigenous societies such as Australian Aborigines and African Bushmen. These "primitive" hunting and gathering societies enjoy material and social plenty without modern technologies. In an essay published three decades ago, "The Original Affluent Society," anthropologist Marshall Sahlins pointed out that these societies work three to five hours a day at an unhurried pace to satisfy their needs. They "keep banker's hours, notably less than modern industrial workers (unionized), who would surely settle for a 21–35 hour week." After a study of their situation, Sahlins concluded: "The world's most primitive people have few possessions, *but they are not poor*."[72] Moreover, they have plenty of time to socialize, sleep, and enjoy life, which is understandable. They do not have to keep up with the barrage of technological advances and software releases that today's high-entropic information societies have to deal with, nor do they have to worry about each technology's peculiar side effects and software's internal bugs.

Regarding technology, some low-entropic advice comes from *U.S. News & World Report*. In an article entitled "Returning to Retrotech," the magazine points out, "Avoiding new gizmos isn't Luddite; in fact, it's often smart." The editors observed that most new high-tech gadgets become obsolete almost instantly, and in a survey they asked the heretical question, Do you need new technology? It was found that many people do not. And they are neither cheapskates nor Luddites. "These people simply believe old technology makes them better thinkers, more efficient workers, and just a little more comfortable." Classic machines, like Leica cameras and DC-3 airplanes, have "a quality of manufacture and design, along with durability." Because today's technologies—whether hardware or software—are invariably complex, the magazine concludes: "Sometimes, the simplicity of retro technology also makes it the best choice."[73]

There is a school of thought that maintains "computers decrease rather than increase complexity."[74] The message that computers simplify our lives in an increasingly complex world has been conveyed by so many people, especially scientists, that it has become accepted as a truth in many circles.[75] But remember that computers are part of our thermodynamic system. Every process and every machine contributes to the increase in the entropy, sometimes called complexity, of the thermodynamic system. Astrophysicists remind us that the universe began simple and has become complex due to increases in entropy. This is also true for our own world. Computers are a major contributor to our current phenomenal increases in complexity and information entropy. Every system that the computer has invaded has become more complex because the computer has increased the entropy of the system. Computers have been around for at least a quarter-century. If they indeed decrease complexity and simplify our lives, we surely would have seen some indications by now.

In the literature, computers and related technologies are often touted not only as fighters of complexity but also solutions to other human problems, including environmental degradation. Computers and communication networks are depicted as tools for achieving a sustainable future for humankind. Even the ecologically minded Voluntary Simplicity movement has adopted this view. In *Voluntary Simplicity*, Duane Elgin, a founder of the movement, writes: "To build a sustainable future, citizens of the earth must see themselves as part of a tightly interdependent system rather than as isolated individuals and nations."[76] How do we achieve this sustainable future and interdependence? Elgin's answer is this:

> The communications revolution plays a critical role in global consciousness raising and consensus building. With the rapid development of sophisticated communication networks, the global consciousness of humanity awakens decisively. The integration of

computers, telephones, television, satellites, and fiber optics into a unified, interactive multimedia system gives the world a powerful voice, and a palpable conscience. The earth has a new vehicle for its collective thinking and invention that transcends any nation or culture. From this communications revolution comes a trailblazing, new level of human creativity, daring, and action in response to the global ecological crisis.[77]

Thermodynamics gives us a different viewpoint, as expressed throughout this book. Every satellite we build and launch contributes to "the global ecological crisis" here on Earth and up in space. Every microchip we build adds to the environmental disorder. Not surprisingly, Silicon Valley now has 29 Superfund sites, making it the densest concentration of highly hazardous waste dumps in the country.[78] Moreover, the region has a serious problem with smog. As researcher Aaron Sachs pointed out in a *World Watch* article, the high-tech communities are not the leafy, green suburban-style environments believed in some quarters. On the contrary, the economics of high-tech development tend to drive out almost everything except the factories and the roads that connect them. In Silicon Valley, traffic congestion is now a way of life. The new computer plants were built on some of the world's richest farmland. Once famous for its immense pear, peach, cherry, and apricot orchards, the Santa Clara Valley is now almost completely paved over.[79] This is hardly a recipe for achieving a sustainable future for humankind, especially in a world where the amount of food-producing land per capita is declining precipitously.

The poor and hungry of the world do not need complex and high-entropic devices, such as computers, which give them information about food. They need low-entropic food. The homeless and the destitute of the world do not need sophisticated Web sites that show them beautiful pictures of expensive homes. They need shelter.

The Laws of Thermodynamics impinge on everything we do. The more things we try to do in a given day, and the more complex machinery we use to accomplish our goal, the more entropic we become. Thus a life-style fostered by "timesaving" products is high-entropic and hectic. A life-style thrust by rapid technological and socioeconomic changes is high-entropic and exhausting. A life-style driven by high rates of production, consumption, and obsolescence is high-entropic and draining.

Chapter 11

The Thermodynamic Imperative

The first and second laws of thermodynamics are of course known to us as well as the Ten Commandments, and probably obeyed more consistently![1]
—H. S. Seifert

Does Science Tell Us How to Live?

In a blistering article entitled "The Superstitions of Science," Leo Tolstoy sharply criticized science for studying "what is most profitable and easy to study" while "neglecting what is more real and important."[2] The Russian novelist and philosopher questioned the value of scientific endeavors: "the chemical constitution of the Milky Way, of the new element helium, . . . of X rays, and the like. But, says the simple common-sense person, none of this is necessary to me; I need to know how to live."[3] He concluded that "science, in order to become science, and to become truly beneficent, and not injurious to mankind . . . must return to the only wise and fruitful understanding of science, according to which its object is the study of how people ought to live."[4]

Many philosophers, including Tolstoy, have looked to science for some social laws of behavior and have been disappointed. It is true that science has not come up with cookbook instructions on how we ought to live, but science does provide us with some general Laws of Nature, especially the First and Second Laws of Thermodynamics. The discipline of thermodynamics, with its concepts of energy and entropy, definitely provides us with a clear-cut choice of "how to live," and at the same time tells us the consequences of choosing a particular way of life.

Since every process or action exhibits an increase in entropy, everything we do leaves a mark in our thermodynamic system, irrevocably. So we should be able to derive a certain code of conduct from the Laws of Thermodynamics. Physicist R. B. Lindsay has proposed that we behave according to the thermodynamic imperative: "while we *do* live we ought always to act in all things in such a way as to produce as much order in our environment as possible. . . . This is the thermodynamic imperative, a normative principle which may serve as the basis for a persuasive ethic in the spirit of the Golden Rule and Kant's categorical imperative."[5]

There is a problem in Lindsay's thermodynamic imperative. As H. S. Seifert has noted, there is no general agreement on "what constitutes maximum order. Some engineers might maintain, for example, that the hours spent by ladies at the hairdressers carefully arranging a coiffure into something called a 'feather cut' or a 'beehive' do not achieve maximum order."[6] Similarly, sociologists may question the intrinsic value of factories that manufacture such highly ordered objects as synthetic diamonds—especially if people are not adequately fed. Besides, all processes including the order-generating varieties increase entropy somewhere in the environment. "Man as a rational being should not blindly create order but instead increase order to improve the quality of life for everyone," writes G. Tyler Miller, Jr. He points out that "blindly crowding the planet with life results in disorder, not order."[7]

We can indeed derive an imperative from thermodynamics because the Laws of Thermodynamics are all-inclusive. "The real imperative that we should learn from thermodynamics is simple and profound: *whenever you do anything, be sure to take into account its present and possible future impact on the surroundings, or environment,*" writes chemist G. Tyler Miller. "This is an ecological imperative that must be learned and applied now if we are to survive or at least to prevent a drastic degradation in the quality of life for all passengers on our fragile craft."[8]

Miller's ecological imperative deals with only part of the thermodynamic equation, the environment. The thermodynamic imperative must also address the system itself, in particular humankind. It should therefore be formulated as follows: *whenever we do anything, we ought to consider its present and possible future impact—entropy—on ourselves, on other human beings, on humanity, on living organisms, and on the environment.*

In the past three centuries, civilization has been largely based on Francis Bacon's guiding principle that knowledge is power. This motto has been repeated in the literature over and over again. Pierre Teilhard de Chardin was an eloquent advocate of this philosophy. He wrote, "*Knowledge for its own sake.* But also, and perhaps still more, *knowledge for power.* . . . *Increased power for increased action.* But, finally and above all, *increased action for increased being.*"[9]

In general, modern civilization has advocated increased knowledge not for increased wisdom but for increased power, action, control, dominance, and conquest. Many physicists and technologists speak about humanity's power over the natural world—an accomplishment of applied physics—but forget to mention the consequences for Nature of applying power. For example, in *The Physicists*, Sir C. P. Snow wrote about the modern achievements of applied physics, especially in the area of nuclear energy, adding: "These results—there are plenty more—come through the physicists' power over the natural world."[10]

What is "power" and what does the exertion of power do? In physics, power is the rate of energy transformations per unit of time. The Second Law tells us that the more energy transformations we perform, the more entropy we generate, and this fact is a fundamental part of our knowledge. The more power we exert on Nature and each other, the more friction and entropy we generate, thus creating a maelstrom of problems for humankind. Today's environmental, social, political, and economic disorders are consequences of all the power we have exerted on Nature and on each other—one person on another, one political group on another, one economic system on another, one nation on another, and one ideological group on another.

We must recognize that we have accumulated a considerable amount of technological capabilities and that the more we use them, the more disorder we are going to create. As physicist Steven Weinberg reminds us, "Technology certainly gives us the power to wreck the environment in which we live."[11] Today we have more degrees of freedom to exert power and generate entropy than ever before. With this freedom must come the responsibility and the wisdom to use power with utmost care.

The Necessity of Projecting a Consistent Scientific Message

Scientists have pointed out that we can no longer remain ignorant of science, because the applications of scientific knowledge profoundly affect our well-being. But as René Dubos has remarked, with few exceptions the layperson knows only a few artifacts and by-products of science. The distinguished biologist was confronted with a fundamental problem: namely, which scientific laws should be emphasized and taught to the general public. He stated that there is no criterion "to guide selection as to what kind of scientific facts would be suitable for the education of the layman."[12] In support of his point of view, in 1970 he wrote:

> When Sir Charles Snow first discussed *The Two Cultures* in his celebrated Rede Lecture of 1959, he stated that an understanding of entropy and of The Second Law of Thermodynamics was the common property of all educated men and an essential part of modern culture. But when he reprinted the same lectures a decade later, he emphasized instead knowledge of molecular biology, the DNA molecule, and its relation to heredity. It is not unlikely that he might still change the priority if he were republishing his lecture today. . . .
>
> Rapid and profound shifts of emphasis have repeatedly occurred in the scientific community, in part because fashions change in science even more than in other types of endeavors, also because social concerns inevitably affect intellectual preoccupations.[13]

Scientists are not alone in projecting a rapidly changing set of ideas. The noted economist Paul Samuelson acknowledged, in the eleventh edition of *Economics* (1980), that economics must recognize the laws of physics. He wrote: "Though economics involves more than technology, its laws must respect those of nature and physics."[14] Samuelson felt that the First and Second Laws of Thermodynamics did indeed deserve respect, since they were consequential to economic activities. However, in subsequent editions with William D. Nordhaus, the Laws of Thermodynamics are never mentioned. What happened?

If the Laws of Thermodynamics were "consequential" to economics in 1980, what made them unimportant a few years later? Nothing. Today the Law of Entropy is even more consequential to economics, as the level of disorder in our socioeconomic system has risen and is still rising. As time elapses and entropy increases, the concept of disorder becomes more worthy of consideration. Inconstant messages by economists not only create confusion and uncertainty but also result in inconsistent economic policies and actions.

These rapid changes in viewpoints and attitudes have added considerable disorder to people's minds. Unfortunately, René Dubos himself changed his position on scientific matters rather drastically, and within a short decade. For example, in *Reason Awake* (1970) he pointed out that countertechnologies, also called technological fixes, usually create new environmental problems of their own, and that we need a new philosophy of humanity and the environment.[15] He wrote:

> Developing *countertechnologies* to correct the new kinds of damage constantly being created by technological innovations is a policy of despair. If we follow this course we shall increasingly behave like hunted creatures, fleeing from one protective device to another, each more costly, more complex, and more undependable than the one before; we shall be concerned chiefly with sheltering ourselves from environmental dangers while sacrificing the values that make life worth living.[16]

But in 1981, in *Celebrations of Life*, Dubos changed his point of view. In a chapter entitled "Optimism, Despite It All," he expressed his strong conviction that "scientific knowledge enables us to learn how to solve most of the practical problems of the modern world—from shortages of food or energy, to environmental degradation and probably even overpopulation."[17]

How would scientific knowledge solve such problems as energy shortage? Obviously, through some technology. Writing $E = mc^2$ on the blackboard or in books does not generate electricity. Eventually we have to develop a technology—a process—to convert mass into energy. This is exactly what the nuclear industry did, and then came the disorders (entropy) of nuclear technology—nuclear waste products, radioactivity, and occasional accidents.

How would scientific knowledge solve problems related to the degradation of the environment? Again, through some technology. Consider the problem of the warming of our planet. Can we develop countertechnologies to alleviate this problem? Yes, so long as we are willing to pay our dues to the Second Law, which is always ready to collect. As it turns out, the technologies that could counteract the greenhouse effect are almost available. But as *Newsweek* has noted, "the cures, worthy of Rube Goldberg, could prove worse than the disease."[18]

We could, for instance, deliberately tamper with airplane engines so that they emit soot into the atmosphere, blocking sunlight; we could use naval guns to fire shells filled with dust into the stratosphere to create sunlight-reflecting layer of particles; we could let loose billions of aluminized, hydrogen-filled balloons into the upper atmosphere, where they would reflect the Sun's rays; and we could orbit tens of thousands of huge mirrors to deflect sunlight.[19]

The problem remains, however, that all these processes are, themselves, rich in entropy production. "Besides cost," wrote *Newsweek*, "the trouble with technological fixes is that they can leave a problem worse than before. Despite pervasive use of pesticides, for instance, more crops than ever are lost to fungus and insects, and the chemicals pollute ground water and leave residues on produce. No one knows the ecological effects of a constant veil of stratospheric dust."[20]

"Technological fixes can turn around and bite you." remarked Jessica Tuchman Mathews, now President of the Carnegie Endowment for International Peace. *Newsweek* added: "That's a lesson that Americans who believe in the magic wand of technology have trouble learning."[21] What we all need to learn is the Second Law of Thermodynamics, which tells us that we cannot remedy the disorders generated by our technological processes by introducing "counter" processes, because the new ones will generate their own peculiar entropies.

We may choose to call a given process a "controlling mechanism," but the Second Law pays no attention to what we call it. It collects its dues from *all processes*, whether they are the "real" or the "controlling" ones. "Nor should we have delusions that if we sneak up on Nature with a smile," writes biologist David Ehrenfeld, "muttering the right charms and incantations—'biological control,' 'natural insecticide,' 'mulch'—we will catch her unaware and in a good mood."[22]

Consider the catalytic converter, a pollution control device hailed by Detroit automakers "as a miracle." The device was added to the automobile two decades or so ago to reduce smog, and indeed it has. The converter breaks down compounds of nitrogen and oxygen from the automobile exhaust that can combine with hydrocarbons, also from the automobile, and be cooked by the Sun rays into smog. Scientists had suspected, however, that catalytic converters sometimes rearrange the nitrogen-oxygen compounds to form nitrous oxide, popularly known as laughing gas. Nitrous oxide is a potent greenhouse gas, about 300 times more potent than carbon dioxide, the most prevalent greenhouse gas.

In the spring of 1998, the Environmental Protection Agency published a study estimating that nitrous oxide now accounts for about 7.2 percent of greenhouse gases, and that automobiles and trucks, most fitted with catalytic converters, generate about half of that nitrous oxide. Moreover, the study showed that nitrous oxide is one of the few greenhouse gases whose emissions are increasing rapidly. This growth comes about from the increase in the distances traveled by vehicles with catalytic converters. Ironically, as improvements were made in catalytic converters, which eliminated more of the smog-producing nitrogen and oxygen compounds, vehicles produced more nitrous oxide. While our air is being cleaned up, our world is being warmed up. As one government official remarked, "the problem created by the converter is classic. 'You've got people trying to solve one problem, and as is not uncommon, they've created another.'"[23]

The Second Law has a one-track mind. It keeps track of one quantity—entropy—and makes sure that this quantity always increases in *all* processes. The Second Law pays no attention to the purpose of a particular technology. Consequently, whether computers are used for education, calculation, dissemination of information, or for military purposes, the Second Law ensures that the disorder in our thermodynamic system increases. Whether satellites are launched for monitoring the depletion of the ozone layer, for communication, or for spying on other countries, the entropy of our environment increases.

Prominent scientists who state that scientific knowledge can solve most environmental degradation problems send the wrong message to the public and to policymakers. In his first State of the Union address, in 1969, President Nixon proposed taking care of environmental problems by mobilizing the energy "of the same reservoir of inventive genius that created those problems in the first place."[24]

The fact remains that the scientific community and humanity at large are checkmated by the Second Law of Thermodynamics. Once entropy is generated, scientists cannot invent an anti-entropy mechanism to reverse the process and repair the damage. Max Planck made it clear that entropy produced by thermodynamic processes cannot be reversed "when the process has once taken place."[25] Not even Maxwell's demon could do it. We can escape from or counteract the effects of other laws of Nature, but there is no way to neutralize the Laws of Thermodynamics.

We are being showered with inconsistent messages. It is not uncommon to pick up a magazine and find incongruous approaches to today's problems, especially those about the environment. Perceptive readers are noticing: "Your April 16 [1990] issue is a perfect illustration of the inconsistencies characterizing the national approach to environmental concerns," observed one *Newsweek* reader. "Your cover highlights the problems of the Mississippi; the story of this abused national resource concludes that the river 'carries with it the sorrows of a continent.' Then your back cover carries a picture of a couple happily spraying their lawn with pesticides/herbicides. I need no lectures about freedom of advertising. But unless and until we all get our respective priorities in order, it's easy to understand why much of the populace views environmentalism as just another phase."[26]

If we all shared the basic scientific knowledge that all processes and technologies increase entropy—in conformity with the Second Law—and applied it consistently in all our activities, we would then have a common basis for discussion and for tackling the problems of modern life.

Making Entropy a Part of Our Daily Language

For some reason, we seldom find the concept of entropy in news-papers, magazines, or even books. And when we do, we rarely find it as an explanation for the increasing disorder of our thermody-namic system. For example, entropy was mentioned when *Newsweek* did a cover story on the importance of scientific knowledge. The article, called "Not Just for Nerds," noted that educators take inquisitive youngsters "and turn them off to science completely and irreversibly," adding, "The one sure way to turn kids off science is to make them memorize vocabulary lists. By one estimate, there are more new words in a science course than in an introductory for-eign-language class."[27] Consequently, in this high-entropic environ-ment we need to emphasize the teaching of scientific concepts rather than scientific facts, which are increasing massively. So the magazine gathered about five dozen "terms" that anyone with a good knowledge of high-school science should be familiar with. Geothermal energy, $E = mc^2$, and entropy were among the chosen terms. *Newsweek* defined entropy succinctly as "the measure of a disorder of a system."[28]

Because the concepts of energy and entropy are related to the First and Second Laws of Thermodynamics, which govern all nat-ural processes, they are "not just for nerds," kids, and science stu-dents. They are for all of us—farmers, educators, business execu-tives, environmentalists, economists, technologists, policymakers, ecologists, philosophers, reporters, writers, and movie producers. While it is important to know the relationship $E = mc^2$, it is much more important to be aware of the Second Law, which states un-equivocally that when a certain mass is converted into energy through nuclear processes, the entropy—the measure of disorder of our thermodynamic system—increases irreversibly. This is true for all other processes—whether aluminum is turned into a beer can or copper is transformed into a microchip.

In our everyday language, we use the word energy but not en-tropy. In books and magazine articles, the word pollution is usually substituted for entropy, especially when the environment is involved. We find a typical statement in *Audubon*, the magazine of the National Audubon Society: "The automobile pollutes, as does virtually every human endeavor, from making a campfire to raising cattle to publishing magazines."[29] A similar statement is found in *U.S. News & World Report*: "At bottom, economic activity generates pollution, whether it is acid rain, toxic waste or smog."[30] Connoisseurs of the Second Law recognize what these assertions mean: that all processes and activities generate entropy. Other words and phrases found in the literature that essentially mean entropy include disorder, waste, complexity, externalities, side effects, collateral effects, hidden costs, and unintended consequences.

On a few occasions, we do find the concepts of entropy and the Second Law described and used. For example, an essay in *Time* pointed out that the Second Law is "one of the most far-reaching commandments of physics."[31] The main ideas behind the Second Law were put forth informally but elegantly in this way:

> The second law defines a quantity called entropy, which is a measure of waste and disorder, and which tends to increase over time inside the beating cylinder of an automobile engine, for example, or inside the universe. . . . The laws of physics say you can never put Alaska back together again once you have dismantled it for its minerals. As a point of national discourse, thermodynamics would be a reminder of mortality and humility. . . . The burden of saving the planet belongs not to technologists but to the rest of us. Sure we can always make better machines, but they will not save us; it is we who have to save them.[32]

As more people—worldwide—become familiar with the Laws of Thermodynamics, especially the Second Law, more of us will be aware of the disorders emanating from our socioeconomic, technological, and intellectual activities, which are increasing massively. Hence, we need to make the concept of entropy an inseparable part of our daily language.

Because thermodynamics came about from the study of heat and heat engines, it has not received the general attention it deserves. The word thermodynamics gives the impression that its laws are only for heat engines and not for everything going on in

Nature. The time has come, however, to regard the First and Second Laws of Thermodynamics as Nature's First and Second Laws— as Nature's Two Commandments.

If we learned from the very beginning of childhood that all processes increase the disorder of our environment, and that all our activities contribute to that disorder, we would be more aware of the consequences of our actions throughout our lives.

Thermodynamics and the Unity of Knowledge

The disorder in knowledge and higher education has increased to such a level that many educators feel it has reached a crisis point. In *The Closing of the American Mind* (1987), Allan Bloom of the University of Chicago decried the state of the American higher education: "The university's evident lack of wholeness in an enterprise that clearly demands it cannot help troubling some of its members."[33] When students arrive at the university, they come across a bewildering variety of departments and courses that do not address one another. The courses are "competing and contradictory, without being aware of it."[34] Unquestionably, human knowledge is in a state of decomposition and diffusion: its disorder—entropy—is noticeably high. Bloom summarized the plight of higher education this way:

> The problem is the lack of any unity of the sciences and the loss of the will or the means even to discuss the issue. The illness above is the cause of the illness below, to which all the good-willed efforts of honest liberal educationists can at best be palliatives. . . .
>
> To repeat, the crisis of liberal education is a reflection of a crisis at the peaks of learning, an incoherence and incompatibility among the first principles with which we interpret the world, an intellectual crisis of the greatest magnitude, which constitutes the crisis of our civilization. But perhaps it would be true to say that the crisis consists not so much in this incoherence but in our inca-

pacity to discuss or even recognize it. Liberal education flourished when it prepared the way for the discussion of a unified view of nature and man's place in it, which the best minds debated on the highest level. It decayed when what lay beyond it were only specialties, the premises of which do not lead to any such vision. The highest is the partial intellect; there is no synopsis.[35]

Bloom strongly believed that the solution to the lack of wholeness in the disciplines lay in the study of the classic Great Books, in which a liberal education incorporates the reading of certain generally recognized classic texts. "One thing is certain," he wrote, "wherever the Great Books make up a central part of the curriculum, the students are excited and satisfied. . . . The advantage they get is an awareness of the classic—particularly important for our innocents; an acquaintance with what big questions were when there were still big questions; models, at the very least, of how to go about answering them; and, perhaps most important of all, a fund of shared experiences and thoughts on which to ground their friendships with one another."[36]

Regrettably, Bloom's solution does not even come close to solving the original problem, which is the lack of unity in the sciences or, generally speaking, the lack of unity in the disciplines. It may be true that wherever Great Books are taught on a massive scale, the "innocent" students have become all excited and satisfied, but the university has certainly not solved the central problem of the lack of wholeness in its enterprise, nor has it made any progress in eliminating the courses that are competing and contradictory. The university has merely provided a new alternative, in a sense a new specialty—which has its own internal inconsistencies and contradictions.

By reading and studying the classic texts, like Plato's *Republic*, students will not learn the fundamental principles of Nature. To find out how Nature works, students must take basic science courses where general Laws of Nature, like the Laws of Thermodynamics, are discussed. Universities must make an effort to teach all students the basic Laws of Nature. Unfortunately, the natural sciences are not synthesized, even though the Laws of Thermodynamics apply to all its branches. The message to science from Sir Julian Huxley is straightforward: "Present-day science needs to synthesize itself before it can be of real service in the necessary task of synthesizing world affairs."[37]

Huxley's plea for the synthesis of scientific knowledge came in the mid-1960s. Similarly, in 1964, Julius Stratton had articulated the need for a synthesis of knowledge as a prerequisite to tackling humanity's problems. The president of MIT had argued that "our efforts must now move to a higher plateau. We can no longer afford to nibble away piece by piece at the problems of the modern city, of transportation, of underdeveloped economies, of automation, or of disarmament. . . . One of the charges that has been most commonly leveled against science is that progress is leading increasingly to the fragmentation of knowledge and the proliferation of a multitude of specialties."[38]

His approach entailed bringing together the many disciplines in a joint effort to solve the ever-increasing problems of modern life: "For in every instance, success will depend upon the joint contributions of physical and biological scientists, of economists and political scientists, of engineers and architects, of historians and philosophers. The task of articulating or welding together these components of learning into systems of understanding offers the highest intellectual challenge of our time."[39] However, bringing together the many scientists, engineers, political scientists, economists, and philosophers generates even more entropy as the disorders—paradoxes, contradictions, inconsistencies—of each discipline permeate each other, creating more heat than light.

And when we speak of such an undertaking, which economists do we have in mind? The Keynesians, supply-siders, monetarists, or all of them? We have to recognize that economists disagree not only on the proper course of action but also on the origin of our economic problems. Similarly, which political scientists do we have in mind? The liberals, neo-liberals, conservatives, middle-of-the-roaders, or all of them? How about technologists and engineers? Which set do we bring in to solve our problems? Those who believe "small is beautiful," or those who believe in "soft" or "hard" technologies? How about the philosophers? Shall we invite the humanists, the existentialists, the pragmatists, the logical positivists, the deconstructionists . . . or all of them?

The idea that the welding together of the disciplines—the socalled multidisciplinary approach—will enable us to solve the problems of the individual disciplines and of modern life is entrenched in the minds of many scientists and educators. But if we make each discipline into a multidisciplinary field and bring the subjective fields into the picture, we would be increasing the complexity and disorder of our knowledge in a big way. In any field of endeavor, introducing additional state variables—especially of the subjective kind—is tantamount to injecting high entropy and uncertainty into it. When problems arise, we will have even more difficulty pin-pointing their sources and making corrections to them because there are too many variables, each one subject to a variety of opin-ions. Under this high-entropic approach, societies will be depen-dent on "multidisciplinary experts" for guidance and leadership. These experts will be propounding their authoritative—but mutu-ally conflicting—opinions, thus generating more confusion and dis-order. Society's intellectual entropy, which is already high, will increase substantially.

To reduce intellectual entropy and confusion we need to do just the opposite. Instead of attempting to analyze every phenomenon and problem in terms of every idea we have come up with—which is a high-entropic activity—we need to find common and general concepts, such as energy and entropy, and use them in all our disci-plines. We will then have a common foundation.

Stratton argued that the endeavor to synthesize knowledge must be initiated at the university, because only there do we find the wide range of interests—from sciences and engineering to arts and humanities—and a common ground for the exchange of ideas. "This is particularly true of an institution of the character of MIT. I do not believe that we can escape the responsibility of taking part in the solution of problems which touch most deeply upon the total welfare of our society. In the synthesis of knowledge, as well as in the creation of new learning, we must lead the way."[40]

Has MIT, or any other university for that matter, taken the lead in synthesizing knowledge? Not at all. While we talk about synthe-sis, we practice analysis, a never-ending process. We exert greater and greater efforts to analyze phenomena and situations with the greatest possible number of variables and perspectives. While we speak of the need for a synthesis of existing concepts and prin-ciples, we practice the art of coupling and networking the various disintegrated and incoherent parts of our knowledge into bigger disordered wholes.

Currently, the "best minds" have a better opportunity to debate—on the highest level—humanity's place in Nature and to discuss a unified view of the world, because science has discovered general Laws of Nature that allow us to view Nature in a unified way—laws that were not known when Aristotle, Shakespeare, Rousseau, Kant, and other great minds were alive. Today's best minds have to make a concentrated effort to study and learn the general Laws of Nature, so that their projected philosophy of life is in unison, and not in contradiction, with these basic truths.

The disorder in higher education reaches beyond the borders of the university system. The graduates become our leaders—our presidents, governors, senators, and representatives—and they steer people in directions that are on a collision course with the Laws of Nature. The graduates become the executives of our corporations, and the chief technologists and architects of tomorrow's world. The graduates become our teachers and they pass on their acquired intellectual entropy to the next generation. The graduates become our economists, our advisors, our writers, and our intellectuals, and they propound ideas that further increase the disorder in knowledge and the world in which we live.

As Huxley has remarked, "mankind is in travail. Our present psychosocial organization is turning into disorganization and is disintegrating."[41] Thus an urgent need exists for a new synthesis of ideas and beliefs, which would act as a supporting framework for a new social system struggling to be born. Science must emerge as the cornerstone of this new synthesis.

As the famous saying goes, "Science and philosophy form a single whole."[42] The saying has come about because originally philosophy and science were unified in purpose, which was, in the words of Plato, "the vision of truth."[43] Until recently, and certainly when Lord Kelvin was alive, basic scientific investigations—that is, unprejudiced, objective, and cooperative studies—were addressed as fundamental questions of natural philosophy.[44]

Basic science has given way to applied science—technology. We are showered daily with news of technological breakthroughs of one kind or another. But technology does not provide us with first principles with which we could interpret the world around us or synthesize our ideas and beliefs. For that we need to look into the pure sciences—that is, go back to the basics and bring forward the general Laws of Nature that govern natural phenomena, including life itself. Huxley remarks that "science so far has given us a vast mass of new knowledge but no knowledge of how to use it. It has indeed become like the sorcerer's apprentice. It has conjured up this dangerous genie we call technology, which is now threatening man's basic ideas about life and how it should be lived."[45]

More recently, distinguished biologist E. O. Wilson of Harvard University has made yet another appeal for the synthesis of knowledge. In his 1998 book *Consilience: The Unity of Knowledge*, he writes:

> During the past thirty years the ideal of the unity of learning, which the Renaissance and Enlightenment bequeathed us, has been largely abandoned. With rare exceptions American universities and colleges have dissolved their curriculum into a slurry of minor disciplines and specialized courses. While the average number of undergraduate courses per institution doubled, the percentage of mandatory courses in general education dropped by more than half. Science was sequestered in the same period; . . . only a third of universities and colleges require students to take at least one course in the natural sciences.[46]

Today, it is possible to receive a Ph.D. in computer science, economics, philosophy, or political science without taking a single course in the natural sciences.

Wilson points out that "Philosophy plays a vital role in intellectual synthesis," adding: "There has never been a better time for collaboration between scientists and philosophers, especially where they meet in the borderlands between biology, the social sciences, and the humanities. We are approaching a new age of synthesis. . . . Philosophy, the contemplation of the unknown, is a shrinking dominion. We have the common goal of turning as much philosophy as possible into science."[47] He believes that through recent advances in biology, primarily in the study of the brain and genetics, we are on the verge of being able to unify the natural sciences, which he assumes are already unified, with the social sciences—the so-called soft sciences—and even the arts, ethics, and religion.

The question remains, What unifies the natural sciences—physics, chemistry, and biology and their sister sciences such as astrophysics and geology? Since the laws of physics appeared first in the universe, we can ask, What unifies physics? As noted earlier, physics has discovered four forces of Nature: gravitational, electromagnetic, strong, and weak forces.

As mentioned in Chapter 4, the dream of physics—and Albert Einstein's dream—is to unify all the fundamental forces of Nature together with all the successful principles of quantum theory and relativity into what has been called the "Theory of Everything." Can this unification explain the behavior of Nature and our universe? Not quite. Something fundamental is still missing, as brought up by astrophysicist Roger Penrose and chemist Ilya Prigogine, among others. They point out that the four forces of Nature and the equations of quantum physics and relativity are time-reversible (with minor exceptions). They make no distinction between future and past. "It seems to me that there are severe discrepancies between what we consciously feel, concerning the flow of time, and what our (marvellously accurate) theories assert about the reality of the physical world," writes Penrose. "Then where in heaven do we look, to find physical laws more in accordance with what we seem to perceive of the world?"[48] As Penrose points out, there is one important place, the same place that Prigogine keeps telling us in book after book. Namely, the Second Law of Thermodynamics.

Thermodynamics has been around for a century and a half, during which many new scientific discoveries, advances, and theories have come about, but nothing has disturbed its axioms, not even the much-heralded and revolutionary quantum theory. As physicist Arnold Sommerfeld has remarked, "In contrast to classical mechanics, thermodynamics has withstood the quantum revolution without having its foundations shaken."[49]

Quantum theory not only did not shake the foundations of thermodynamics, it received help from thermodynamics in the explanation of the irreversibility of measurements. In his celebrated book *Quantum Theory*, David Bohm remarks that "a measurement process is irreversible in the sense that, after it has occurred," reestablishment of the initial conditions of the system's variables "is overwhelmingly unlikely." Bohm then makes the connection to thermodynamics: "This irreversibility greatly resembles that which appears in thermodynamic processes, where a decrease of entropy is also an overwhelmingly unlikely possibility." After a short examination of the situation, he concludes "thermodynamic irreversibility enters into the quantum theory in an integral way."[50]

In fact, thermodynamics and thermodynamic irreversibility enter into every process occurring in Nature, including quantum-mechanical. That is why—if we look hard enough—we find such books as *Entropy for Biologists*, *Thermodynamics for Geologists*, *The Entropy Law and the Economic Process*, and *Entropy and the Unity of Knowledge*.[51] We find the Laws of Thermodynamics and thermodynamic concepts in books on such diverse topics as ecology and information theory. Entropy plays a significant role in the fields of communication and cybernetics.[52] Thermodynamics, of course, is an integral part of chemistry. Chemists know that chemical processes increase the entropy of the thermodynamic system, although chemical firms do not advertise this fact.

We find thermodynamics and its laws in many disciplines because energy and entropy are fundamental to all natural processes, including human activities. And because heat is everywhere and has been present in our universe since its inception. Heat (temperature) is an important state parameter. If our body temperature rises or drops just a few degrees, we invariably feel its effects. Too much body heat or lack of it can cause death. That is why physicians monitor our body temperature when we get sick.

The temperature of the environment is also an important variable, especially for biological systems. Today there is much concern that our planet is warming up, partly due to our massive emission of greenhouse gases and destruction of tropical forests. We are now feeling the effects of heat. Global warming has attracted the attention of a great number of scientists, concerned individuals, and even world leaders. It has also created friction among various ideological groups. How much temperature change are we talking about? As it turns out, the global average temperature has risen 1 degree Celsius in the past 130 years and continues to rise, having picked up some momentum.[53] You would think that such a small change in temperature would be inconsequential. This is certainly not the case, as we are all finding out. We cannot afford to neglect heat and thermodynamics, Max Planck's favorite subject, or the Second Law of Thermodynamics, Planck's dissertation topic.

Planck was so "deeply impressed" by the importance of the Second Law that he continued his studies of entropy, which he regarded "as next to energy the most important property of physical systems."[54] And as events transpired, his perseverance paid off. When blackbody radiation was discovered and nobody could explain the phenomenon, his previous studies of the Second Law of Thermodynamics stood him "in good stead." While all other physicists, including many famous ones, directed their efforts solely on showing the dependence of the intensity of radiation on the temperature, Planck suspected that the basic connection lies in the dependence of entropy on energy. "As the significance of the concept of entropy had not yet come to be fully appreciated," recounts Planck, "nobody paid any attention to the method adopted by me, and I could work out my calculations completely at my leisure, with absolute thoroughness, without fear of interference or competition."[55]

Planck used Boltzmann's entropy relation to arrive at the result that in the microscopic world, energy "can only take on *discrete* energy values" in contrast to classical theory where energy can be continuous. "We now say," Planck wrote, that energy is "quantized." With this simple statement, in 1900, he started a revolution and a new epoch in physical science.[56]

Quantum mechanics has been a boon for technological advances, especially in microelectronics. However, the equations of quantum mechanics are reversible. We cannot use quantum mechanics, relativity theory, or nuclear physics to arrive at a philosophy of life or a worldview. Reversible equations give the incorrect impression that processes occurring in Nature can go either way, that a human being can grow old, then become young again, and repeat the process almost ad infinitum. It is the Second Law of Thermodynamics that states this is impossible. It also tells us we cannot undo what we are doing to our environment, to our ecosystem, and to our economies.

Planck attained a special place in the hearts of many physicists, including Einstein. In 1918, Einstein delivered a brief address on the occasion of Planck's sixtieth birthday, in which he recognized three types of scientists. In the temple of science, he said, there are those who "take to science out of a joyful sense of their superior intellectual power." For them, science is a kind of "sport" that satisfies their personal ambition. Many others enter the temple of science "for purely utilitarian purposes. Were an angel of the Lord to come and drive all the people belonging to these two categories out of the temple," a few people would be left, including Planck—"and that is why we love him."[57] The philosopher Walter Kaufmann calls this third type, the few like Planck, "visionaries."[58]

Planck's vision of the significance of the Second Law has certainly come true. The concept of entropy has solved physical problems even in areas of investigations where physicists thought it inappropriate. Here is a case in point: For the longest time, black holes were in the camp of classical general relativity. No one would dare apply the idea of entropy to black holes, except a brave graduate student at Princeton University—Jacob Bekenstein. He put forward the audacious suggestion that black holes might have entropy, and a tremendous amount of it. According to Brian Greene, "Bekenstein was motivated by the venerable and well-tested *second law of thermodynamics*, which declares that the entropy of a system always increases: Everything tends toward greater disorder."[59] Bekenstein then came up with the equation of the entropy of a black hole mentioned in Chapter 3.

Stephen Hawking also had come up with the same equation, but he had not associated it with entropy—and for a fundamental reason. Black holes, by definition, are black; they do not emit any radiation. If we want to identify a property of black hole with entropy, we need to assign a *temperature* to a black hole; because if there is no temperature, there is no entropy. But if we give a black hole a nonzero temperature, however small, then it has to emit radiation based on well-established physical principles. Then it is no longer a black hole. Hence the impasse.

Hawking bet against the Second Law and rejected Bekenstein's suggestion. He argued that if matter carrying entropy falls into a black hole, this entropy is simply lost. So much for the Second Law. In 1974, however, Hawking made a turnabout. After lengthy and arduous theoretical calculations, he pronounced that black holes are not completely black. They have temperature and entropy. Moreover, the entropy coincided with Bekenstein's formula.[60]

The concept of entropy, although physical in nature, also has a logical counterpart. As discussed in Chapter 3, when the entropy of a system increases so does our ignorance, our uncertainty about its internal arrangements. This tells us that when economies, eco-systems, technologies, or world situations change massively and rapidly, as they are now, our uncertainty and ignorance of our thermodynamic world increase accordingly.

Despite all the advances in scientific endeavors, there is disunity in knowledge. "Clearly something is missing in the way we are educating our children," remarks Frederick Turner, Founders Professor of Arts and Humanities at the University of Texas at Dallas. "What is that missing something? Most fundamentally, perhaps, it is a sense of cognitive unity, a unity which imparts meaning to the world and from which our values unfold." He contends that the academic curriculum, as currently shaped, is "one great obstacle" to this unity. He points out that the scientific and intellectual progress of the last four centuries contain a "gigantic paradox. Every great advance, every profound insight in the sciences and other intellectual disciplines, has torn down the barriers and distinctions between those disciplines; and yet the institutional result of each of these achievements has been the further fragmentation and specialization of the academy."[61]

The discoveries of the Laws of Thermodynamics are excellent examples of what Turner is talking about. The laws and their subsequent application clearly unified the world of physics, chemistry, and biology. Even before these were discovered, Newton unified mechanics, astronomy, and optics. His mathematical formulation of the laws of physics also bridged the border between mathematics and physics. From Newton on, mathematics became the language of physics. In many instances, especially in physics, Nature speaks to us in the language of mathematics. In the words of Sir James Jeans, physicist and astronomer, "the Great Architect seems to be a mathematician."[62]

Frederick Turner has a simple proposal. He advocates using the universe and its progression as a model for our educational system. The universe began with an enormous heat in the order of 10^{32} degrees, and with a few fundamental laws of physics present. There was no chemistry going on because there can be no chemistry above 3,000 degrees. After a few hundred thousand years of expansion, the universe cooled sufficiently to allow the formation of stable atoms. Biology had to wait much longer because it requires much cooler temperatures than chemistry. Humankind appeared recently in the universe, and all the human sciences from anthropology to zoology to political science and economics have a history of but a few thousand years. All biological processes obey the laws of biology, which obey the laws of chemistry, which in turn obey the laws of physics, which require some mathematics to understand.

"The general structure of the hierarchy of the universe is now fairly clear," Turner writes. Therefore, it is essential that we change our basic thinking. The academy needs a major restructuring. Students must first learn some mathematics so they will have the background to learn physics. After learning the major principles of physics, they will learn the important laws of chemistry, then of biology, and finally they will study other fields of interest. As Turner points out, "a detailed knowledge of the fields that underlie one's own discipline is not necessary, as long as we are able to understand their major principles and laws, their most powerful theoretical generalizations."[63]

In physics, the most powerful generalizations are contained in the Laws of Thermodynamics. All the forces of Nature in their interactions and all natural processes obey the Laws of Energy and Entropy. The Laws of Thermodynamics can be used by educators as a focal point for the unification of knowledge.

We can also use the Laws of Thermodynamics to derive a practical philosophy of life as expressed throughout this book. A high-entropic life-style based on high rates of production, consumption, and obsolescence exacts a price from us. The Laws of Thermodynamics tell us that the more chemicals and technologies we introduce in our world and the more power we exert on Nature, the more disorder we generate in our environment, which in turn creates a variety of problems for humanity.

We need a new generation of philosophers who would penetrate into the basic laws of science, especially the Laws of Thermodynamics, and make an effort to illuminate scientific truths and present a synthesized view of Nature. We need a new generation of teachers and parents who would explain to the young the concepts of order and disorder, and teach values that are in conformity with the Laws of Nature. We need a new generation of reporters, writers, and communicators who would not only report environmental and socioeconomic disorders but would also explain observed phenomena objectively in terms of natural laws. We need a new generation of economists who would propose socioeconomic policies that are in harmony with the Laws of Nature. We need a new generation of world leaders who are familiar with the Laws of Thermodynamics, who share a common understanding of how Nature works, and who would work together to reduce the expansion of world disorder.

In essence, we need a new generation of people who are bonded together with a common understanding of the basic Laws of Thermodynamics, and who would work in concert to tackle humanity's outstanding problems. The thermodynamic clock is ticking—*irreversibly!*

Notes

Introduction

1. Quote by Carver Mead, in George Gilder, *Microcosm: The Quantum Revolution in Economics and Technology* (New York: Simon and Schuster, 1989), p. 11.
2. Al Gore, *Earth in the Balance: Ecology and the Human Spirit* (Boston: Houghton Mifflin, 1992), p. 186.
3. René Dubos, *Reason Awake: Science for Man* (New York: Columbia University Press, 1970), p. xi.

Chapter 1. Nature's First Law

1. R. Bruce Lindsay, "The Concept of Energy and Its Early Historical Development," in R. Bruce Lindsay, ed., *Energy: Historical Development of the Concept* (Stroudsburg, Penn.: Dowden, Hutchinson & Ross, 1975), p. 13.
2. Donald W. Rogers, "An Informal History of the First Law of Thermodynamics," *Chemistry*, December 1976, p. 12.
3. Ibid.
4. Morton Mott-Smith, *The Concept of Energy Simply Explained* (New York: Dover Publications, 1964), p. 63.
5. Arnold Sommerfeld, *Thermodynamics and Statistical Mechanics, Lectures on Theoretical Physics*, Vol. V, trans. by J. Kestin (New York: Academic Press, 1956), p. 1.
6. Antoine Laurent Lavoisier, "Introduction of the Term 'Caloric' for the Substance of Heat," in Lindsay, *Energy*, p. 204.
7. Rogers, "An Informal History," p. 11.
8. Lavoisier, "Introduction of the Term 'Caloric,'" p. 204.
9. Robert Boyle, "The Nature of Heat," in Lindsay, *Energy*, p. 182. *See also* Stanley W. Angrist and Loren G. Hepler, *Order and Chaos: Laws of Energy and Entropy* (New York: Basic Books, 1967), p. 59.
10. Benjamin Thompson (Count Rumford), "Source of Heat from Friction," in Lindsay, *Energy*, p. 224.

11. Mott-Smith, *Energy Simply Explained*, p. 82.

12. "Editor's Comments on Papers 28 through 31," in Lindsay, *Energy*, p. 257.

13. Mott-Smith, *Energy Simply Explained*, p. 84.

14. Henry A. Bent, *The Second Law: An Introduction to Classical and Statistical Thermodynamics* (New York: Oxford University Press, 1965), p. 15.

15. Ibid., p. 14.

16. D. S. L. Cardwell, *From Watt to Clausius: The Rise of Thermodynamics in the Early Industrial Age* (Ithaca, N.Y.: Cornell University Press, 1971), p. 238.

17. Mott-Smith, *Energy Simply Explained*, p. 113.

18. Ibid.

19. R. Bruce Lindsay, *Men of Physics: Julius Robert Mayer—Prophet of Energy* (Oxford, Eng.: Pergamon Press, 1973).

20. Max Planck, *Treatise on Thermodynamics*, 3rd ed., trans. by Alexander Ogg (New York: Dover Publications, 1945), p. 40.

21. Edward Teller, *Energy from Heaven and Earth* (San Francisco: W. H. Freeman, 1979), p. 309.

Chapter 2. Nature's Second Law

1. C. P. Snow, *The Two Cultures and the Scientific Revolution* (New York: Cambridge University Press, 1959), pp. 15–16.

2. S. Carnot, "Reflections on the Motive Power of Fire, and on Machines Fitted to Develop That Power," in Joseph Kestin, ed., *The Second Law of Thermodynamics* (Stroudsburg, Penn.: Dowden, Hutchinson & Ross, 1976), pp. 16–35.

3. D. S. L. Cardwell, *From Watt to Clausius: The Rise of Thermodynamics in the Early Industrial Age* (Ithaca, N.Y.: Cornell University Press, 1971), p. 212.

4. William Thomson (Lord Kelvin), "On an Absolute Thermometric Scale Founded on Carnot's Theory of the Motive Power of Heat, and Calculated from Regnault's Observations," in Kestin, *The Second Law of Thermodynamics*, p. 54.

5. Carnot, "Reflections on Motive Power of Fire," p. 33.

6. Thomson, "On an Absolute Thermometric Scale," p. 56.

7. Quoted in Cardwell, *From Watt to Clausius*, p. 235.

8. Thomson, "On an Absolute Thermometric Scale," p. 54.

9. Quoted in Cardwell, *From Watt to Clausius*, p. 244.

10. Ibid., p. 241.

11. Rudolf Clausius, "On the Moving Force of Heat, and the Laws Regarding the Nature of Heat Itself Which Are Deducible Therefrom,"

in Kestin, *The Second Law of Thermodynamics*, p. 90.

12. William Thomson (Lord Kelvin), "On the Dynamical Theory of Heat, with Numerical Results Deduced from Mr. Joule's Equivalent of a Thermal Unit, and M. Regnault's Observations on Steam," in Kestin, *The Second Law of Thermodynamics*, p. 121.

13. Rudolf Clausius, "On the Application of the Theorem of the Equivalence of Transformations to the Internal Work of a Mass of Matter," in Kestin, *The Second Law of Thermodynamics*, p. 133.

14. Max Planck, *Treatise on Thermodynamics*, 3rd ed., trans. by Alexander Ogg (New York: Dover Publications, 1945), p. 89.

15. Ludwig Boltzmann, *Lectures on Gas Theory*, trans. by Stephen G. Brush (Berkeley: University of California Press, 1964), pp. 444–45.

16. Arnold Sommerfeld, *Thermodynamics and Statistical Mechanics, Lectures on Theoretical Physics*, Vol. V, trans. by J. Kestin (New York: Academic Press, 1956), p. 19.

17. Planck, *Treatise on Thermodynamics*, pp. 82–83.

18. Ibid., p. 88.

19. G. J. Whitrow, *The Nature of Time* (New York: Holt, Rinehart and Winston, 1973, copyright 1972 by Thames and Hudson Ltd., London), p. 83.

20. Planck, *Treatise on Thermodynamics*, p. 86.

21. Mortimer J. Adler and Charles Van Doren, eds., *Great Treasury of Western Thought: A Compendium of Important Statements on Man and His Institutions by the Great Thinkers in Western History* (New York: R. R. Bowker, 1977), p. 1128.

22. Arthur S. Eddington, *The Nature of the Physical World* (New York: The Macmillan Company, 1929), p. 103.

23. Francis Weston Sears, *Principles of Physics I: Mechanics, Heat, and Sound*, 2nd ed. (Reading, Mass.: Addison-Wesley, 1950), p. 459.

24. Rudolf Clausius, "On Different Forms of the Fundamental Equations of the Mechanical Theory of Heat and Their Convenience for Application," in Kestin, *The Second Law of Thermodynamics*, pp. 186–87.

25. Ibid., p. 193.

26. Quoted in G. Tyler Miller, Jr., *Energetics, Kinetics, and Life: An Ecological Approach* (Belmont, Calif.: Wadsworth Publishing Company, 1971), p. 143.

27. George Musser, "Taming Maxwell's Demon," *Scientific American*, February 1999, p. 24.

28. Quoted in D. ter Haar, *Elements of Statistical Mechanics* (New York: Rinehart & Company, 1954), p. 161.

29. Quoted in Leon Brillouin, *Science and Information Theory*, 2nd ed. (New York: Academic Press, 1962), p. 168.

30. Richard P. Feynman, Robert B. Leighton, and Matthew Sands, *The

Feynman Lectures on Physics, Vol. I (Reading, Mass.: Addison-Wesley, 1963), pp. 46-2 to 46-4.

31. Planck, *Treatise on Thermodynamics*, p. 106.
32. Quoted in Dennis E. Hensley, "The Time of Your Life," *Writer's Digest*, August 1982, p. 22.
33. A. Cornelius Benjamin, "Ideas of Time in the History of Philosophy," in J. T. Fraser, ed., *The Voices of Time: A Cooperative Survey of Man's Views of Time as Expressed by the Sciences and by the Humanities* (New York: George Braziller, 1966), pp. 3–30.
34. Hans Reichenbach, *The Direction of Time* (Berkeley: University of California Press, 1956), p. 16.
35. Quoted in Benjamin, "Ideas of Time in the History of Philosophy," p. 18.
36. Quoted in Peter V. Coveney and Roger Highfield, *The Arrow of Time: A Voyage Through Science to Solve Time's Greatest Mystery* (New York: Fawcett Columbine, 1991, copyright 1990), p. 82.
37. Whitrow, *The Nature of Time*, p. 122.
38. Quoted in ibid., pp. 125–26.
39. Martin J. Rees, *Before the Beginning: Our Universe and Others* (Reading, Mass.: Addison-Wesley, 1997), p. 211. *See also* p. 156.
40. "Translator's Introduction," in Boltzmann, *Lectures on Gas Theory*, p. 13.
41. Boltzmann, *Lectures on Gas Theory*, p. 216.
42. "Translator's Introduction," in ibid., p. 14.
43. Boltzmann, *Lectures on Gas Theory*, p. 215.
44. Feynman, Leighton, and Sands, *Feynman Lectures*, p. 46-7.
45. Richard Feynman, *The Character of Physical Law* (Cambridge, Mass.: The MIT Press, 1967), p. 171.
46. Ibid., p. 156.

Chapter 3. Nature's Laws in Action

1. G. Tyler Miller, Jr., *Energetics, Kinetics, and Life: An Ecological Approach* (Belmont, Calif.: Wadsworth Publishing Company, 1971), p. 200.
2. George Porter, "The Laws of Disorder, Part 4, Equilibrium—The Limit of Disorder," *Chemistry*, September 1968, p. 16.
3. Miller, *Energetics, Kinetics, and Life*, p. 29.
4. Jacob Bekenstein, "Black Holes and Entropy," *Phys. Rev.*, D7, 1972, pp. 2333–46; Stephen W. Hawking, "Particle Creation by Black Holes," *Commun. Math. Phys.*, 43, 1975, pp. 199–220.
5. Roger Penrose, *The Emperor's New Mind: Concerning Computers, Minds, and the Laws of Physics* (New York: Penguin Books, 1991), p. 341.
6. Charles Kittel and Herbert Kroemer, *Thermal Physics*, 2nd ed. (San Francisco: W. H. Freeman, 1980), p. 47.

7. Bert Murray, in "Pepper . . . and Salt," *Wall Street Journal*, 11 July 1984, p. 25.
8. "Matters of Scale: Visible vs. Invisible Waste," *World Watch*, November/December 1999, p. 37.
9. Willard Van Orman Quine, "On What There Is," in Hans Regnéll, ed., *Readings in Analytical Philosophy* (Stockholm: Läromedelsförlagen, 1971), p. 38.
10. Quoted in Robert H. Romer, *Energy: An Introduction to Physics* (San Francisco: W. H. Freeman, 1976), p. 197.
11. William Thomson (Lord Kelvin), "On a Universal Tendency in Nature to the Dissipation of Mechanical Energy," in Joseph Kestin, ed., *The Second Law of Thermodynamics* (Stroudsburg, Penn.: Dowden, Hutchinson & Ross, 1976), p. 197.
12. Craig J. Hogan, *The Little Book of the Big Bang: A Cosmic Primer* (New York: Copernicus, 1998), p. 25.
13. Freeman J. Dyson, "Energy in the Universe," *Scientific American*, September 1971, pp. 50–59.
14. Paul Davies, *The Runaway Universe* (New York: Harper & Row, 1978), p. 162.
15. Martin J. Rees, *Before the Beginning: Our Universe and Others* (Reading, Mass.: Addison-Wesley, 1997), p. 47. *See also* Marcus Chown, *Afterglow of Creation: From the Fireball to the Discovery of Cosmic Ripples* (Sausalito, Calif.: University Science Books, 1996), pp. 21–34.
16. Joseph Silk, *A Short History of the Universe* (New York: Scientific American Library, 1997), p. 45. *See also* p. 53.
17. Penrose, *The Emperor's New Mind*, p. 326.
18. Ibid., p. 339.
19. Hogan, *The Little Book of the Big Bang*, p. 157.
20. Craig J. Hogan, Robert P. Kirshner, and Nicholas B. Suntzeff, "Surveying Space-time with Supernovae," *Scientific American*, January 1999, p. 46.
21. The Editors, "Special Report: Revolution in Cosmology," *Scientific American*, January 1999, p. 45.
22. Quoted in Claude E. Shannon and Warren Weaver, *The Mathematical Theory of Communication* (Urbana, Ill.: University of Illinois Press, 1964, copyright 1949), p. 3.
23. Leo Szilard, Z. *Physik*, 53, 1929, p. 840.
24. D. ter Haar, *Elements of Statistical Mechanics* (New York: Rinehart & Company, 1954), p. 160.
25. Lyric Wallwork Winik, "Before the Next Epidemic Strikes," *Parade Magazine*, 8 February 1998, p. 6.
26. Lindsey Grant, *Juggernaut: Growth on a Finite Planet* (Santa Ana, Calif.:

Seven Locks Press, 1996), p. 14.

27. Fred Setterberg and Lonny Shavelson, *Toxic Nation: The Fight to Save Our Communities from Chemical Contamination* (New York: John Wiley, 1993), p. 28.
28. Grant, *Juggernaut*, p. 15.
29. Silk, *A Short History of the Universe*, p. 82.
30. Stuart Clark, *Towards the Edge of the Universe: A Review of Modern Cosmology* (New York/Chichester, U.K.: John Wiley/Praxis, 1997), pp. 77–79.
31. Marc Lappé, *Evolutionary Medicine: Rethinking the Origins of Disease* (San Francisco: Sierra Club Books, 1994), p. 51.
32. Ilya Prigogine, *From Being to Becoming: Time and Complexity in the Physical Sciences* (San Francisco: W. H. Freeman, 1980), pp. 5–6.
33. Stanley W. Angrist and Loren G. Hepler, *Order and Chaos: Laws of Energy and Entropy* (New York: Basic Books, 1967), pp. 130–32.
34. Fritjof Capra, *The Turning Point: Science, Society, and the Rising Culture* (New York: Simon and Schuster/Bantam Books, 1982), pp. 269–72.
35. Nicholas Coni, William Davison, and Stephen Webster, *Ageing: The Facts* (Oxford: Oxford University Press, 1984), p. 40.
36. Erwin Schrödinger, *What is Life?* (Cambridge: Cambridge University Press, 1944), p. 72.
37. Ibid., pp. 74–75.
38. Henry A. Bent, *The Second Law: An Introduction to Classical and Statistical Thermodynamics* (New York: Oxford University Press, 1965), p. 138.
39. Ilya Prigogine, *The End of Certainty: Time, Chaos, and the New Laws of Nature* (New York: The Free Press, 1997), p. 63.
40. Quoted in Nancy Shute, "Why Do We Age?" *U.S. News & World Report*, 18/25 August 1997, p. 57.
41. Max Planck, *Scientific Autobiography, and Other Papers,* trans. by Frank Gaynor (New York: Greenwood Press, 1968, copyright 1949 by Philosophical Library), p. 17.
42. "How People Will Live to Be 100 or More," *U.S. News & World Report*, 4 July 1983, p. 73.
43. Shute, "Why Do We Age?" p. 55.
44. Bernard L. Strehler, *Time, Cells, and Aging*, 2nd ed. (New York: Academic Press, 1977), p. 13.
45. Ibid., p. 10.
46. Max Planck, *Treatise on Thermodynamics*, 3rd ed., trans. by Alexander Ogg (New York: Dover Publications, 1945), p. 84.
47. Prigogine, *From Being to Becoming*, p. 212.
48. Lawrence E. Lamb, M.D., *Get Ready for Immortality* (New York: Harper & Row, 1974); Alvin Silverstein, *Conquest of Death* (New York: Macmillan, 1979).

49. M. Scott Peck, M.D., *The Road Less Traveled: A New Psychology of Love, Traditional Values and Spiritual Growth* (New York: Simon and Schuster/ Touchstone Book, 1978), p. 264.
50. Ibid., pp. 264–65.
51. Miller, *Energetics, Kinetics, and Life*, pp. 139–43.

Chapter 4. Entropy in Human Knowledge

1. Ruth Nanda Anshen, "World Perspectives—*What This Series Means*," in Werner Heisenberg, *Physics and Beyond: Encounters and Conversations*, trans. by Arnold J. Pomerans, Vol. 42, *World Perspectives* (New York: Harper & Row, 1971), p. xv.
2. T. S. Eliot, *Complete Poems and Plays* (New York: Harcourt, Brace, 1952), p. 96.
3. Keith R. Symon, *Mechanics*, 2nd ed. (Reading, Mass.: Addison-Wesley, 1960), pp. 2–3.
4. Edward O. Wilson, *Consilience: The Unity of Knowledge* (New York: Alfred A. Knopf, 1998), p. 39.
5. Gerald Holton, "The Organization of Scientific Work Among the Sciences and in Relation to Technology and Culture," in *Science and Synthesis: An International Colloquium Organized by Unesco on the Tenth Anniversary of the Death of Albert Einstein and Teilhard de Chardin* (New York: Springer-Verlag, 1971, copyright 1967), pp. 158–59.
6. Richard Feynman, *The Character of Physical Law* (Cambridge, Mass.: The MIT Press, 1967), p. 39.
7. Steven Weinberg, *Dreams of a Final Theory* (New York: Pantheon Books, 1992), p. 218.
8. Brian Greene, *The Elegant Universe: Superstrings, Hidden Dimensions, and the Quest for the Ultimate Theory* (New York: W.W. Norton & Company, 1999), p. 19.
9. R. A. De Millo, R. J. Lipton, and A. J. Perlis, "Social Processes and Proofs of Theorems and Programs," *Comm. of the ACM*, May 1979, p. 272.
10. Steven Weinberg, "Five and a Half Utopias," *Atlantic Monthly*, January 2000, p. 113.
11. Quoted in George E. Jones with Carey W. English, "Social Sciences: Why Doubts Are Spreading Now," *U.S. News & World Report*, 31 May 1982, p. 71.
12. William Greider, "The Education of David Stockman," *Atlantic Monthly*, December 1981, pp. 27–54.
13. "A Conversation with S. E. Luria: Consequences of America's 'Ignorance of Science,'" *U.S. News & World Report*, 14 May 1984, p. 76.
14. Quoted in Gary Chapman, "Well-Informed Citizens Increasingly Rare

in Information Age," *Los Angeles Times*, 17 July 2000, p. C3.

15. Ibid.
16. "A Conversation with S. E. Luria," p. 76.
17. Joseph Weizenbaum, *Computer Power and Human Reason: From Judgment to Calculation* (San Francisco: W. H. Freeman, 1976), pp. 277–78.
18. "A Conversation with S. E. Luria," p. 76.
19. Quoted in Porter G. Perrin, *Writer's Guide and Index to English*, 3rd ed. (Chicago, Ill.: Scott, Foresman, 1959), p. 25.
20. Sharon Begley, "Don't Believe What You Read," *Newsweek*, 6 August 1990, p. 71.
21. Ibid.
22. "Tempests in a Test Tube: Two New Studies Ask Why Scientists Cheat," *Newsweek*, 2 February 1987, p. 64.
23. Ibid.
24. Ibid.
25. William Broad and Nicholas Wade, *Betrayers of the Truth: Fraud and Deceit in the Halls of Science* (New York: Simon and Schuster, 1982); Alexander Kohn, *False Prophets: Fraud and Error in Science and Medicine* (New York: Basil Blackwell, 1986).
26. Kohn, *False Prophets*, p. 2.
27. Julian L. Simon, *The Ultimate Resource 2*, rev. ed. (Princeton, N.J.: Princeton University Press, 1996), p. 81.
28. George Gilder, *Wealth and Poverty* (New York: Basic Books, 1981), p. 260.
29. Ibid., p. 261.
30. Mentioned in "Playboy Interview: George Gilder," *Playboy*, August 1981, p. 69.
31. R. Buckminster Fuller, *Critical Path* (New York: St. Martin's Press, 1981), p. 109.
32. Ibid., p. 361.
33. Ibid., book jacket and p. xi.
34. Ibid., p. xxxi.
35. William Oliver Martin, *The Order and Integration of Knowledge* (Ann Arbor: The University of Michigan Press, 1957), p. vii.
36. Pierre Teilhard de Chardin, *The Future of Man* (New York: Harper & Row, 1964), p. 133.

Chapter 5. The United States in High Entropy

1. Duncan Caldwell, *Reader's Digest*, May 1949, p. 7.
2. Ralph Keyes, *Timelock: How Life Got So Hectic and What You Can Do About It* (New York: HarperCollins Publishers, 1991), p. 8.
3. A. Kent MacDougall, "Stressful Times; Americans: Life in the Fast

Lane," *Los Angeles Times*, 17 April 1983, p. 1.

4. Nancy Gibbs, "Cover Story: How America Has Run Out of Time," *Time*, 24 April 1989, p. 59.
5. Quoted in H. V. Routh, *Money, Morals and Manners as Revealed in Modern Literature* (London: Ivor Nicholson and Watson, 1935), p. 67.
6. Quoted in Keyes, *Timelock*, p. 183.
7. Ibid., p. 82.
8. Ibid., p. 83.
9. Quoted in A. Kent MacDougall, "Overload: More Time Is Less Time," *Los Angeles Times*, 18 April 1983, p. 1.
10. Routh, *Money, Morals and Manners as Revealed in Modern Literature*, p. 67.
11. MacDougall, "More Time Is Less Time," p. 15.
12. James Dale Davidson, *The Squeeze* (New York: Summit Books, 1980), pp. 105–20.
13. MacDougall, "More Time Is Less Time," p. 14.
14. Staffan Burenstam Linder, *The Harried Leisure Class* (New York: Columbia University Press, 1970), p. 46.
15. Keyes, *Timelock*, p. 85.
16. Quoted in René Dubos, *Celebrations of Life* (New York: McGraw-Hill, 1981), p. 218.
17. Lindsey Grant, *Juggernaut: Growth on a Finite Planet* (Santa Ana, Calif.: Seven Locks Press, 1996), pp. 87–88.
18. Susan Headden, "Cover Story: The Junk Mail Deluge," *U.S. News & World Report*, 8 December 1997, pp. 40–41.
19. Alvin Toffler, *Future Shock* (New York: Random House/Bantam Books, 1970), p. 157.
20. John L. Kirkley, "Too Much to Read; or, Stop the Presses, I Want to Get Off," *Datamation*, December 1981, p. 27.
21. Alvin Toffler, *The Third Wave* (New York: William Morrow/Bantam Books, 1980), p. 189.
22. Peter J. Denning, "ACM President's Letter, Electronic Junk," *Comm. of the ACM*, March 1982, pp. 163–65.
23. Lewis Mumford, *The Myth of the Machine: The Pentagon of Power* (New York: Harcourt Brace Jovanovich, 1970), p. 293.
24. John Foley, "Infoglut," *Informationweek*, 30 October 1995, p. 32.
25. Bertrand Russell, *In Praise of Idleness; and Other Essays* (New York: Simon and Schuster, 1972), pp. 9–29.
26. "Flying Through Airline Loopholes," *U.S. News & World Report*, 31 July 1989, p. 61.
27. MacDougall, "More Time Is Less Time," p. 14.
28. Keyes, *Timelock*, p. 65.

29. Jane Bryant Quinn, "The Virtues of Simplicity," *Newsweek*, 1 February 1999, p. 33.
30. "Working 9 to 5—on Vacation," *Los Angeles Times*, 6 September 1998, p. L3.
31. MacDougall, "Americans: Life in the Fast Lane," p. 1.
32. Quoted in Shannon Brownlee and Matthew Miller, "Lies Parents Tell Themselves About Why They Work," *U.S. News & World Report*, 12 May 1997, p. 60.
33. Juliet B. Schor, *The Overspent American: Upscaling, Downshifting, and the New Consumer* (New York: Basic Books, 1998), p. 20.
34. Martha Shirk, Neil G. Bennett, and J. Lawrence Aber, *Lives on the Line: American Families and the Struggle to Make Ends Meet* (Boulder, Colo.: Westview Press, 1999), p. 2.
35. Phillip J. Longman, "The Cost of Children," *U.S. News & World Report*, 30 March 1998, p. 51.
36. Arlie Russell Hochschild, *The Time Bind: When Work Becomes Home and Home Becomes Work* (New York: Metropolitan Books, 1997), p. 6.
37. Brownlee and Miller, "Lies Parents Tell Themselves About Why They Work," p. 61.
38. Steve Lopez, "Hide and Seek," *Time*, 11 May 1998, p. 60.
39. Hochschild, *The Time Bind*, p. 50.
40. Stacey Schultz, "The Youngest Victims: Where Are the Simple Joys of Childhood?" *U.S. News & World Report*, 8 March 1999, p. 63.
41. Joannie M. Schrof and Stacey Schultz, "Melancholy Nation," *U.S. News & World Report*, 8 March 1999, p. 56.
42. David Elkind, *The Hurried Child: Growing Up Too Fast Too Soon* (Reading, Mass.: Addison-Wesley, 1981).
43. David Elkind, *Ties That Stress: The New Family Imbalance* (Cambridge, Mass.: Harvard University Press, 1994), pp. 188–89, 201.
44. "Civic Groups List Threats to Children," *Los Angeles Times*, 29 November 1999, p. A13.
45. Andrew Vachss, "Our Endangered Species," *Parade Magazine*, 29 March 1998, p. 1.
46. Quoted in René Dubos, *Reason Awake: Science for Man* (New York: Columbia University Press, 1970), p. 179.
47. "A Conversation with Mary Eleanor Clark: 'We Have Become Servants,' Not Masters, of Technology," *U.S. News & World Report*, 17 August 1981, p. 48.
48. Toffler, *The Third Wave*, p. 246.
49. Gibbs, "Cover Story: How America Has Run Out of Time," p. 60.
50. Toffler, *The Third Wave*, p. 253.
51. Karen Nussbaum, "Employer, You're No James Bond," *Playboy*, April 1990, p. 56.

52. Ibid.
53. Rodger Doyle, "Privacy in the Workplace," *Scientific American*, January 1999, p. 36.
54. "Fast Times," *48 Hours*, CBS News, 8 March 1990.
55. Andrea Gabor, "On-the-Job Straining," *U.S. News & World Report*, 21 May 1990, p. 53.
56. Stuart Silverstein, "Tracking Life in the Fast Lane," *Los Angeles Times*, 30 April 1990, p. A1.
57. Gabor, "On-the-Job Straining," p. 51.
58. Katie Hafner, "Are Customers Ever Right? Service's Decline and Fall," *New York Times*, 20 July 2000, p. D8.
59. Quoted in Ibid.

Chapter 6. The Agricultural-Industrial Complex

1. John E. Thompson, Letters to the Editor: "Topsoil: Our Renewable Black Gold," *Wall Street Journal*, 7 December 1981, p. 27.
2. Lester R. Brown, *Tough Choices: Facing the Challenge of Food Scarcity* (New York: W.W. Norton & Company, 1996).
3. Lester R. Brown, "Human Food Production as a Process in the Biosphere," *Scientific American*, September 1970, pp. 163–64.
4. Earl Cook, *Man, Energy, Society* (San Francisco: W. H. Freeman, 1976), p. 155.
5. Sandra Postel, *Pillar of Sand: Can the Irrigation Miracle Last?* (New York: W.W. Norton & Company, 1999), p. 6.
6. Ibid.
7. Paul Hawken, *The Ecology of Commerce: A Declaration of Sustainability* (New York: HarperCollins Publishers, 1993), p. 3.
8. Sandra Postel, *Last Oasis: Facing Water Scarcity* (New York: W.W. Norton & Company, 1992, 1997), p. 54.
9. Brown, *Tough Choices*, p. 78.
10. Ibid., p. 65.
11. Postel, *Last Oasis*, pp. 52–53.
12. Ibid., pp. 54–55.
13. Lester R. Brown, "Feeding Nine Billion," in Lester R. Brown et al., *State of the World 1999* (New York: W.W. Norton & Company, 1999), p. 127.
14. Brown, *Tough Choices*, p. 27.
15. Evan Eisenberg, *The Ecology of Eden* (New York: Alfred A. Knopf, 1998), p. 31.
16. Gary Gardner, "Recycling Organic Wastes," in Lester R. Brown et al., *State of the World 1998* (New York: W.W. Norton & Company, 1998), p. 100.

17. Ibid.
18. Clive A. Edwards, "The Impact of Pesticides on the Environment," in David Pimentel and Hugh Lehman, eds., *The Pesticide Question: Environment, Economics, and Ethics* (New York: Chapman & Hall, 1993), p. 13.
19. Mark L. Winston, *Nature Wars: People vs. Pests* (Cambridge, Mass.: Harvard University Press, 1997), p. 11.
20. Ibid.
21. Brown, "Human Food Production," p. 169.
22. Robert van den Bosch, *The Pesticide Conspiracy* (Garden City, N.Y.: Doubleday, 1978), pp. 24–25.
23. David Pimentel et al., "Environmental and Economic Impacts of Reducing U.S. Agricultural Pesticide Use," in Pimentel and Lehman, *The Pesticide Question*, p. 226.
24. David Pimentel et al., "Assessment of Environmental and Economic Impacts of Pesticide Use," in *The Pesticide Question*, p. 57.
25. Theo Colborn, Dianne Dumanoski, and John Peterson Myers, *Our Stolen Future: Are We Threatening Our Fertility, Intelligence, and Survival?—A Scientific Detective Story* (New York: Dutton Books, 1996), pp. 122–141.
26. Ibid.; *see also* Betsy Carpenter, "Investigating the Next 'Silent Spring,'" *U.S. News & World Report*, 11 March 1996, p. 50.
27. World Health Organization cited in Winston, *Nature Wars*, p. 12.
28. Ibid.
29. Ibid., p. 13.
30. Ibid., pp. 13, 16–17.
31. Daniel J. Hillel, *Out of the Earth: Civilization and the Life of the Soil* (New York: The Free Press, 1991), p. 3.
32. Quoted in James Risser, "A Renewed Threat of Soil Erosion: It's Worse Than the Dust Bowl," *Smithsonian*, March 1981, p. 130.
33. David Pimentel et al., "Environmental and Economic Costs of Soil Erosion and Conservation Benefits," *Science*, 24 February 1995, p. 1117.
34. Al Gore, *Earth in the Balance: Ecology and the Human Spirit* (Boston: Houghton Mifflin, 1992), p. 3.
35. Ibid.
36. Pimentel et al., "Environmental and Economic Costs of Soil Erosion and Conservation Benefits," p. 1117.
37. Gardner, "Recycling Organic Wastes," p. 96.
38. Ross Gelbspan, *The Heat Is On: The High Stakes Battle Over Earth's Threatened Climate* (New York: Addison-Wesley, 1997), p. 155.
39. G. Tyler Miller, Jr., *Energetics, Kinetics, and Life: An Ecological Approach* (Belmont, Calif.: Wadsworth Publishing Company, 1971), p. 300.

Chapter 7. What Does the Second Law Really Say?

1. Max Planck, *Scientific Autobiography, and Other Papers*, trans. by Frank Gaynor (New York: Greenwood Press, 1968, copyright 1949 by Philosophical Library), p. 18.
2. Henry Adams, *The Degradation of the Democratic Dogma* (New York: The Macmillan Company, 1919), pp. 216, 230.
3. Max Planck, *Treatise on Thermodynamics*, 3rd ed., trans. by Alexander Ogg (New York: Dover Publications, 1945), pp. 103–4.
4. Donella H. Meadows et al., *The Limits to Growth* (New York: Universe Books, 1972); *The Global 2000 Report to the President* (New York: Penguin Books, 1982).
5. René Dubos, *Celebrations of Life* (New York: McGraw-Hill, 1981), p. 142.
6. Julian L. Simon, *The Ultimate Resource* (Princeton, N.J.: Princeton University Press, 1981), p. 40.
7. H. E. Goeller and Alvin M. Weinberg, "The Age of Substitutability," *Science*, 20 February 1976, p. 684.
8. J. Peter Vajk, *Doomsday Has Been Cancelled* (Culver City, Calif.: Peace Press, 1978), p. 51.
9. Simon, *The Ultimate Resource*, p. 89.
10. Julian L. Simon and Herman Kahn, "Introduction," in Julian L. Simon and Herman Kahn, eds., *The Resourceful Earth: A Response to Global 2000* (New York: Basil Blackwell, 1984), p. 1.
11. Ibid., p. 25.
12. R. B. Lindsay, "Entropy Consumption and Values in Physical Science," *American Scientist*, Autumn-September 1959, pp. 378–80.
13. Edward Teller, *Energy from Heaven and Earth* (San Francisco: W. H. Freeman, 1979), p. 9.
14. Ibid., p. 221.
15. Dubos, *Celebrations of Life*, p. 249.
16. Planck, *Scientific Autobiography, and Other Papers*, p. 37.
17. Planck, *Treatise on Thermodynamics*, p. 82.
18. Arnold Sommerfeld, *Thermodynamics and Statistical Mechanics, Lectures on Theoretical Physics*, Vol. V, trans. by J. Kestin (New York: Academic Press, 1956), p. 38.
19. Ibid.
20. Abraham Pais, *Niels Bohr's Times: In Physics, Philosophy, and Polity* (New York: Oxford University Press, 1991), p. 161.
21. Ibid.
22. William Thomson (Lord Kelvin), "On a Universal Tendency in Nature to the Dissipation of Mechanical Energy," in Joseph Kestin, ed., *The Second Law of Thermodynamics* (Stroudsburg, Penn.: Dowden, Hutchin-

son & Ross, 1976), pp. 194–97.

23. Ibid., p. 194.
24. Teller, *Energy From Heaven and Earth*, p. 91.
25. "No Free Lunch: A New Look at Solar Energy," *Time*, 10 January 1983, p. 53.
26. Robert H. Romer, *Energy: An Introduction to Physics* (San Francisco: W. H. Freeman, 1976), p. 362.
27. Sommerfeld, *Thermodynamics and Statistical Mechanics*, p. 40.
28. Robert Emden, "Why Do We Have Winter Heating?" *Nature*, 21 May 1938, p. 909.
29. Chris Bright, "Anticipating Environmental 'Surprise,'" in Lester R. Brown et al., *State of the World 2000* (New York: W.W. Norton & Company, 2000), p. 29.
30. Teller, *Energy From Heaven and Earth*, p. 133.
31. Michael D. Lemonick, "Global Warming: Feeling the Heat," *Time*, 2 January 1989, pp. 36–37.
32. Ibid., p. 36.
33. Lester R. Brown and Christopher Flavin, "A New Economy for a New Century," in Lester R. Brown et al., *State of the World 1999* (New York: W.W. Norton & Company, 1999), p. 14. *See also* "Trends Jumping Off the Charts," *Worldwatch News Alert*, <http://www.worldwatch.org/alerts/990530.html>, viewed 7 June 1999.
34. Michael D. Lemonick, "Courting Disaster," *Time*, 3 November 1997, pp. 64–68; Michael Satchell, "Islands Fear a Rising Tide," *U.S. News & World Report*, 27 September 1999, p. 43; Bright, "Anticipating Environmental 'Surprise,'" pp. 29–32. *See also* Charles J. Hanley, "Global Warming Talks Go Slowly as Apparent Effects Speed Up," *Los Angeles Times*, 28 September 1997, p. A2.
35. Lemonick, "Global Warming: Feeling the Heat," p. 37.
36. Seth Dunn and Christopher Flavin, "Destructive Storms Drive Insurance Losses Up," *Worldwatch News Alert*, <www.worldwatch.org/alerts/990325.html>, viewed 7 June 1999.
37. Seth Dunn, "Weather Damages Drop," in Lester R. Brown et al., *Vital Signs 2000* (New York: W.W. Norton & Company, 2000), p. 77.
38. Janet N. Abramovitz and Seth Dunn, "Record Year for Weather-Related Disasters," *Worldwatch News Alert*, <http://www.worldwatch.org/alerts/981127.html>, viewed 15 December 1998; "The Winds and Waters Came," *U.S. News & World Report*, 28 December 1998/4 January 1999, p. 83.
39. Jim Loney, "5 Big Hurricanes, but Fewer Deaths," *USA Today*, 29 November 1999, p. 4A.
40. Sharon Begley, "Too Much Hot Air," *Newsweek*, 20 October 1997, p. 49.

41. Quoted in Karl Deutsch, "Some Problems of Science and Values," in John G. Burke, ed., *The New Technology and Human Values* (Belmont, Calif.: Wadsworth Publishing Company, 1966), p. 37.

42. Michael W. Miller, "Findings of Toxin Leakage in Silicon Valley Hurt Chip Makers' Reputation for Safety," *Wall Street Journal*, 29 August 1984, p. 25.

43. Ibid.

44. "Benchmarks: Don't Drink the Water," *Datamation*, 1 June 1986, p. 60.

45. Aaron Sachs, "Virtual Ecology," *World Watch*, January/February 1999, p. 19.

46. Michael Rogers with Richard Sandza, "Technology: Trouble in the Valley," *Newsweek*, 25 February 1985, p. 93.

47. Joseph LaDou, M.D., "The Not-So-Clean Business of Making Chips," *Technology Review*, May-June 1984, pp. 22, 34.

48. Sachs, "Virtual Ecology," p. 16.

49. "Electric 'Hash' New Menace of High-Tech Age," *U.S. News & World Report*, 13 February 1984, p. 62.

50. Robert DeMatteo, *Terminal Shock: The Health Hazards of Video Display Terminals*, 2nd ed. (Toronto: NC Press, 1986); Marilyn K. Goldhaber, Michael R. Polen, and Robert A. Hiatt, "The Risk of Miscarriage and Birth Defects Among Women Who Use Visual Display Terminals During Pregnancy," *American Journal of Industrial Medicine*, 13, 1988, pp. 695–706.

51. Edward Tenner, *Why Things Bite Back: Technology and the Revenge of Unintended Consequences* (New York: Alfred A. Knopf, 1996), p. 172.

52. Robert O. Becker, M.D., *Cross Currents: The Promise of Electromedicine, The Perils of Electropollution* (Los Angeles: Jeremy P. Tarcher, distributed by St. Martin's Press, 1990), p. xiv.

53. Quoted in Ilya Prigogine, *The End of Certainty: Time, Chaos, and the New Laws of Nature* (New York: The Free Press, 1997), p. 154.

54. John Robison, "Appendix: The Articles, Steam and Steam-Engines," in R. Bruce Lindsay, ed., *The Control of Energy* (Stroudsburg, Penn.: Dowden, Hutchinson & Ross, 1977), pp. 152–54.

55. Norbert Wiener, *Cybernetics, or Control and Communication in the Animal and the Machine* (New York: John Wiley, 1948).

56. André-Marie Ampère, "Introduction of the Word Cybernetics," in Lindsay, *The Control of Energy*, p. 323.

57. Quoted in Robert Caussin, "Ampère and Cybernetics," in ibid., pp. 325–26.

58. John McPhee, *The Control of Nature* (New York: Farrar, Straus and Giroux, 1989), book jacket.

59. Stephen Budiansky, "It's Time to Begin the Blame Sweepstakes," *U.S.*

News & World Report, 4 June 1990, p. 11.

60. R. Bruce Lindsay, "Editor's Comments on Papers 22 Through 26," in Lindsay, *The Control of Energy*, p. 322.

61. Marc Lappé, "Antibiotic Roulette," *Reader's Digest*, October 1981, pp. 226–28.

62. J. Madeleine Nash, "Attack of the Superbugs," *Time*, 31 August 1992, p. 62.

63. Ibid.

64. Lyric Wallwork Winik, "Before the Next Epidemic Strikes," *Parade Magazine*, 8 February 1998, pp. 6–7.

65. Marc Lappé, *Evolutionary Medicine: Rethinking the Origins of Disease* (San Francisco: Sierra Club Books, 1994), p. 8.

66. Hendrik Hertzberg, "Summer's Bloodsuckers," *Time*, 10 August 1992, p. 46.

67. Ellen Ruppel Shell, "Resurgence of a Deadly Disease," *Atlantic Monthly*, August 1997, p. 48.

68. Jeffrey Kluger, "Mosquitoes Get Deadly," *Time*, 8 September 1997, p. 69.

69. Nancy Shute, "Pills Don't Come with a Seal of Approval," *U.S. News & World Report*, 29 September 1997, p. 74.

70. Ampère, "Introduction of the Word Cybernetics," p. 323.

71. Henry Margenau, *The Nature of Physical Reality: A Philosophy of Modern Physics* (Woodbridge, Conn.: Ox Bow Press, 1977 Reprint), p. 212.

Chapter 8. Economics, the Environment, and the Laws of Thermodynamics

1. Nicholas Georgescu-Roegen, *Energy and Economic Myths: Institutional and Analytical Economic Essays* (New York: Pergamon Press, 1976), p. 54.

2. Lester C. Thurow, *Dangerous Currents: The State of Economics* (New York: Random House, 1983), p. 12.

3. "The 'Whoops' Bubble Bursts," *Newsweek*, 8 August 1983, p. 61.

4. Robert L. Heilbroner and Lester C. Thurow, *Economics Explained: Everything You Need to Know About How the Economy Works and Where It's Going*, newly rev. and updated (New York: Simon & Schuster, 1998), p. 41.

5. "Economics: A Guide to Understanding the Supply-Siders," *Business Week*, 22 December 1980, p. 76.

6. Ibid., p. 77.

7. Milton Friedman, "Painless Revenue," *Newsweek*, 5 April 1982, p. 63.

8. Alvin Toffler, *The Third Wave* (New York: William Morrow/Bantam Books, 1980), p. 140.

9. Margaret Loeb, "Class Machines: Computers May Widen Gap in School

Quality Between Rich and Poor," *Wall Street Journal*, 26 May 1983, p. 1.

10. Ashley Dunn, Greg Miller, and Charles Piller, "U.S., Firms Overreacted to Y2K Fix, Critics Say," *Los Angeles Times*, 2 January 2000, p. A1.

11. "Faith and Freedom Must Be Our Guiding Stars," Full Text of Reagan Address, *Los Angeles Herald Examiner*, 7 February 1985, p. A19.

12. Dr. Brian Gould, "The Promise and Problem of Medical Technology," *Consumer Health Advocate*, Blue Cross of California, Fall 1989, p. 2.

13. Heilbroner and Thurow, *Economics Explained*, p. 67.

14. Quoted in Betsy Morris, "Latest Tuition Increases Outstrip Inflation as Prestige Colleges Struggle to Catch Up," *Wall Street Journal*, 24 September 1981, Section 2, p. 27.

15. Al Gore, *Earth in the Balance: Ecology and the Human Spirit* (Boston: Houghton Mifflin, 1992), pp. 189, 347.

16. Gary Gardner and Payal Sampat, "Forging a Sustainable Materials Economy," in Lester R. Brown et al., *State of the World 1999* (New York: W.W. Norton & Company, 1999), p. 42.

17. Heilbroner and Thurow, *Economics Explained*, p. 157.

18. Ibid.

19. Quoted in Philip Elmer-Dewitt, "Planet of the Year: Nuclear Power Plots a Comeback," *Time*, 2 January 1989, p. 41.

20. "A Scorecard of Government Waste," *Newsweek*, 14 August 1989, p. 20.

21. Burr Leonard, "Cleaning up," *Forbes*, 1 June 1987, p. 52.

22. "Matters of Scale: Visible vs. Invisible Waste," *World Watch*, November/December 1999, p. 37.

23. Gardner and Sampat, "Forging a Sustainable Materials Economy," p. 49.

24. Julie Marquis, "Smog Study of Children Yields Ominous Results," *Los Angeles Times*, 18 March 1999, p. A1.

25. Richard P. Turco, *Earth Under Siege: From Air Pollution to Global Change* (New York: Oxford University Press, 1997), p. 141.

26. Mark Hertsgaard, "Our Real China Problem," *Atlantic Monthly*, November 1997, p. 97.

27. "Many Major Rivers Are Polluted, Report Says," *Los Angeles Times*, 30 November 1999, p. A10.

28. Chris Bright, "Anticipating Environmental 'Surprise,'" in Lester R. Brown et al., *State of the World 2000* (New York: W.W. Norton & Company, 2000), p. 34. *See also* Mark Hertsgaard, *Earth Odyssey: Around the World in Search of Our Environmental Future* (New York: Broadway Books, 1998), p. 169.

29. Marvin S. Soroos, *The Endangered Atmosphere: Preserving a Global Commons* (Columbia, S.C.: University of South Carolina Press, 1997), pp. 6–9, 65–67.

30. Bernard J. Nebel and Richard T. Wright, *Environmental Science: The Way the World Works*, 6th ed. (Upper Saddle River, N.J.: Prentice-Hall, 1998), p. 404.

31. Ibid., pp. 403–5. *See also* Soroos, *The Endangered Atmosphere*, p. 6.

32. Turco, *Earth Under Siege*, p. 281.

33. Gareth Porter and Janet Welsh Brown, *Global Environmental Politics*, 2nd ed. (Boulder, Colo.: Westview Press, 1996); Jacqueline Vaughn Switzer with Gary Bryner, *Environmental Politics: Domestic and Global Dimensions*, 2nd ed. (New York: St. Martin's Press, 1998).

34. Turco, *Earth Under Siege*, p. 260.

35. Ibid., pp. 173–74.

36. "Our Dirty Air," *U.S. News & World Report*, 12 June 1989, p. 51. *See also* Dan Fagin, "Despite Control Efforts, Acid-Rain Effects Continue to Worsen," *Los Angeles Times*, 24 October 1999, p. A4.

37. Charles E. Little, *The Dying of the Trees: The Pandemic in America's Forests* (New York: Viking, 1995), p. 45.

38. Ibid., pp. ix–x.

39. Fred Setterberg and Lonny Shavelson, *Toxic Nation: The Fight to Save Our Communities from Chemical Contamination* (New York: John Wiley, 1993), p. 94.

40. Ibid., pp. 2–3.

41. Turco, *Earth Under Siege*, p. 317.

42. Edward O. Wilson, *The Diversity of Life* (Cambridge, Mass.: Belknap Press of Harvard University Press, 1992).

43. John Tuxill and Chris Bright, "Losing Strands in the Web of Life," in Lester R. Brown et al., *State of the World 1998* (New York: W.W. Norton & Company, 1998), p. 41.

44. Ibid.

45. Peter M. Vitousek et al., "Human Domination of Earth's Ecosystems," *Science*, 25 July 1997, pp. 494–99.

46. Gregg Easterbrook, "Cleaning Up; Part 5: The Ecosphere," *Newsweek*, 24 July 1989, p. 42.

47. Gregg Easterbrook, "Cleaning Up; Part 2: Water Pollution: Visible Results," *Newsweek*, 24 July 1989, p. 36.

48. Sharon Begley with Mary Hager, "Science: Keep Holding Your Breath," *Newsweek*, 4 June 1990, p. 68.

49. Ibid.

50. "Methanol: Panacea with Problems," *Time*, 18 September 1989, p. 18.

51. Turco, *Earth Under Siege*, p. 179.

52. "Fill'er Up with Methyl," *Newsweek*, 1 May 1989, p. 67.

53. Turco, *Earth Under Siege*, p. 179.

54. Eduardo Gentil, "Brazil's Alcohol-Car Binge Dries Up," *Wall Street*

Journal, 15 October 1981, p. 27.

55. Jillian Bailey, "Bus Plan Approved Over Objections," *Los Angeles Independent*, 4 March 1998, p. 1.

56. "Every Day Is Earth Day with Nuclear Energy," *Newsweek*, 16 April 1990, Special Advertising Section.

57. John Langone, "Planet of the Year: Waste—A Stinking Mess," *Time*, 2 January 1989, p. 45.

58. Ibid., p. 44.

59. Richard J. Barnet and John Cavanagh, *Global Dreams: Imperial Corporations and the New World Order* (New York: Simon & Schuster, 1994), p. 289; Center for Investigative Reporting and Bill Moyers, *Global Dumping Ground: The International Traffic in Hazardous Waste* (Washington: Seven Locks Press, 1990), p. 17.

60. Center for Investigative Reporting and Moyers, *Global Dumping Ground*, p. vii.

61. Wassily Leontief, "Letters: Academic Economics," *Science*, 9 July 1982, p. 104.

62. Paul A. Samuelson, *Economics*, 10th ed. (New York: McGraw-Hill, 1976), p. 6.

63. Vermont Royster, "Thinking Things Over: A Little Humility, Please," *Wall Street Journal*, 1 June 1983, p. 28.

64. Thurow, *Dangerous Currents*, p. 236.

65. Ibid., p. 21.

66. Milton Friedman, *Bright Promises, Dismal Performance: An Economist's Protest* (New York: Harcourt Brace Jovanovich, 1983).

67. Eliot Janeway, "Making Monkeys Out of Economists and Bureaucrats," *Los Angeles Times* (Book Review), 29 May 1983, p. 3.

68. Leontief, "Letters: Academic Economics," pp. 104–7.

69. Lindley H. Clark, Jr., "Everyone Asks the Wrong Economic Question," *Wall Street Journal*, 19 January 1982, p. 23.

70. Georgescu-Roegen, *Energy and Economic Myths*, p. 53.

71. Ibid.

72. Ibid.

73. Ibid.

74. Paul A. Samuelson, *Economics*, 11th ed. (New York: McGraw-Hill, 1980), p. 747.

75. Ibid.

76. James Dale Davidson, *The Squeeze* (New York: Summit Books, 1980), pp. 226–27.

77. Samuelson, *Economics*, 11th ed., p. 747.

78. Ibid.

79. Davidson, *The Squeeze*, p. 226.

80. "Smothering the Waters," *Newsweek*, 10 April 1989, pp. 54–57.
81. Ibid.
82. John G. Mitchell, "In the Wake of the Spill," *National Geographic*, March 1999, p. 106.
83. Ibid., pp. 101–6, 111.
84. Edward Tenner, *Why Things Bite Back: Technology and the Revenge of Unintended Consequences* (New York: Alfred A. Knopf, 1996), pp. 88–89.
85. Raveendra N. Batra, *The Myth of Free Trade: A Plan for America's Economic Revival* (New York: Charles Scribner's Sons, 1993), p. 223.
86. Nicholas Georgescu-Roegen, *The Entropy Law and the Economic Process* (Cambridge, Mass.: Harvard University Press, 1971).
87. Herman E. Daly, *Beyond Growth: The Economics of Sustainable Development* (Boston: Beacon Press, 1996), pp. 197–98.
88. Ibid., p. 198.
89. Max Planck, *Scientific Autobiography, and Other Papers*, trans. by Frank Gaynor (New York: Greenwood Press, 1968, copyright 1949 by Philosophical Library), pp. 33–34.
90. Michael Prowse, "Save Planet Earth from Economists," *Financial Times*, 10 February 1992, p. A-28.
91. K. C. Cole, *The Universe and the Teacup: The Mathematics of Truth and Beauty* (New York: Harcourt Brace, 1998), p. 46.
92. Growth from "BEA News Release," *Business of Economic Analysis*, U.S. Department of Commerce, <http://www.bea.doc.gov/bea/newsrel/gdp199f.htm>, viewed 30 June 1999; emergency shelter requests from David Holmstrom, "Where Will I Sleep Tonight?" *Christian Science Monitor*, 3 February 1999, p. 11.
93. Holmstrom, "Where Will I Sleep Tonight?" p. 11.
94. Quoted in Jessie Halladay, "Demand Up for Services for the Poor," *USA Today*, 14 December 2000, p. 21A.
95. Hertsgaard, "Our Real China Problem," p. 102.

Chapter 9. Why Things Look So Good on the Horizon—Until We Get There

1. Quoted in "Thoughts on the Business of Life," *Forbes*, 3 January 1983, p. 300.
2. Peter Stoler, "Energy: Pulling the Nuclear Plug," *Time*, 13 February 1984, p. 34.
3. Quoted in ibid., p. 35.
4. Ibid., p. 34.
5. Michael Freemantle, "Ten Years After Chernobyl Consequences Are Still Emerging," *Chemical & Engineering News*, 29 April 1996, pp. 18–28;

"10 Years After Chernobyl," *U.S. News Online*, <http://www.usnews.com/usnews/issue/cherno.htm>, viewed 28 April 1998.

6. Stoler, "Energy: Pulling the Nuclear Plug," p. 34.

7. Quoted in René Dubos, *Reason Awake: Science for Man* (New York: Columbia University Press, 1970), p. 95.

8. Edward Tenner, *Why Things Bite Back: Technology and the Revenge of Unintended Consequences* (New York: Alfred A. Knopf, 1996), book jacket and p. xi.

9. "A Conversation with Val L. Fitch, 'We Are in Danger of Losing Our Scientific Leadership,'" *U.S. News & World Report*, 21 June 1982, p. 56.

10. "A Conversation with Morris Kline, Mathematics: From Precision to Doubt in 100 Years," *U.S. News & World Report*, 26 January 1981, p. 63.

11. B. F. Skinner, *Walden Two* (New York: The Macmillan Company, 1962, copyright 1948), p. 14.

12. Keller Breland and Marian Breland, "The Misbehavior of Organisms," *American Psychologist*, November 1961, pp. 681–84.

13. Ibid., p. 681.

14. Ibid., p. 683.

15. James V. McConnell, "Criminals Can Be Brainwashed—Now," *Psychology Today*, April 1970, p. 74.

16. José M. R. Delgado, M.D., *Physical Control of the Mind: Toward a Psychocivilized Society*, Vol. 41, *World Perspectives* (New York: Harper & Row, 1969), p. 223.

17. Ibid., p. 14.

18. Ibid., p. 35.

19. Ibid., p. 179.

20. Ibid., p. 141.

21. Ibid., pp. 259–60.

22. Paul Brodeur, *Currents of Death: Power Lines, Computer Terminals, and the Attempt to Cover Up Their Threat to Your Health* (New York: Simon and Schuster, 1989); Robert O. Becker, M.D., *Cross Currents: The Promise of Electromedicine, The Perils of Electropollution* (Los Angeles: Jeremy P. Tarcher, distributed by St. Martin's Press, 1990).

23. "A Conversation with Isaac Bashevis Singer: 'Modern Man Remains the Wildest Animal,'" *U.S. News & World Report*, 19 December 1983, p. 54.

24. B. F. Skinner, "Freedom and the Control of Men," *The American Scholar*, 25, 1955–56, pp. 47–65, reprinted in John G. Burke, ed., *The New Technology and Human Values* (Belmont, Calif.: Wadsworth Publishing Company, 1966), p. 283.

25. "See Dick and Jane Lie, Cheat, and Steal: Teaching Morality to Kids," hosted by Tom Selleck, KTLA Channel 5, Los Angeles, Calif., 23 April 1989.

26. J. Robert Oppenheimer, "Einstein's Presence," in *Science and Synthesis: An International Colloquium* Organized by Unesco on the Tenth Anniversary of the Death of Albert Einstein and Teilhard de Chardin (New York: Springer-Verlag, 1971, copyright 1967), p. 8.
27. C. P. Snow, *The Physicists* (Boston: Little, Brown, 1981), p. 16.
28. Edward Teller, *Energy from Heaven and Earth* (San Francisco: W. H. Freeman, 1979), p. 218.
29. Charles W. Petit, "A Fresh Jolt for Fusion," *U.S. News & World Report*, 28 September 1998, p. 61.
30. Robert H. Romer, *Energy: An Introduction to Physics* (San Francisco: W. H. Freeman, 1976), p. 492.
31. Ibid., p. 487.
32. David Rittenhouse Inglis, "Guest Editorial: The Energy Frontier," in G. Tyler Miller, Jr., *Energy and Environment: The Four Energy Crises*, 2nd ed. (Belmont, Calif.: Wadsworth Publishing Company, 1980), p. 112.
33. Teller, *Energy from Heaven and Earth*, p. 212.
34. Ibid., p. 218.
35. G. Tyler Miller, Jr., *Energetics, Kinetics, and Life: An Ecological Approach* (Belmont, Calif.: Wadsworth Publishing Company, 1971), pp. 298–99.
36. Gerard K. O'Neill, *2081: A Hopeful View of the Human Future* (New York: Simon and Schuster, 1981), p. 62.
37. Katherine Rizzo, "Price of a Space Vacation? Astronomical!" *Los Angeles Times*, 19 April 1998, p. A26.
38. Jeffrey Kluger, "Who Needs This?" *Time*, 23 November 1998, p. 90.
39. Ibid., pp. 90–91.
40. James R. Chiles, "Casting a High-Tech Net for Space Trash," *Smithsonian*, January 1999, pp. 46–55.
41. Ibid.
42. "A Conversation with Gerard O'Neill, 2081: 'Routine Trips into Space' for 200 Million People a Year," *U.S. News & World Report*, 3 August 1981, pp. 60–61.
43. O'Neill, *2081*, p. 63.
44. Miller, *Energetics, Kinetics, and Life*, p. 299.
45. Ibid.
46. Joel E. Cohen and David Tilman, "Biosphere 2 and Biodiversity: The Lessons So Far," *Science*, 15 November 1996, p. 1151.
47. Art Pine, "Interdependence: It's Not Entirely a Plus," *Wall Street Journal*, 27 June 1983, p. 1.
48. Ibid.
49. John Maynard Keynes, "National Self-Sufficiency," *Yale Review*, XXII, June 1933, Vol. 4, pp. 755–69.
50. Pine, "Interdependence: It's Not Entirely a Plus," p. 1.

51. Chris Bright, *Life Out of Bounds: Bioinvasion in a Borderless* World (New York: W.W. Norton & Company, 1998), pp. 13–14. *See also* Joseph B. Verrengia, "Alien Species' Invasions Leave No Continent Untouched," *Los Angeles Times*, 24 October 1999, p. A37.
52. Quoted in Herman E. Daly and John B. Cobb, Jr., *For the Common Good: Redirecting the Economy Toward Community, the Environment, and a Sustainable Future*, 2nd ed., updated and expanded (Boston: Beacon Press, 1994), p. 209.
53. John M. Culbertson, *International Trade and the Future of the West* (Madison, Wisc.: Twenty-First Century Press, 1984).
54. Daly and Cobb, *For the Common Good*, p. 211.
55. Herman E. Daly, *Beyond Growth: The Economics of Sustainable Development* (Boston: Beacon Press, 1996), p. 145.
56. William Greider, *One World, Ready or Not: The Manic Logic of Global Capitalism* (New York: Simon & Schuster, 1997), p. 11.
57. Paul Hawken, *The Ecology of Commerce: A Declaration of Sustainability* (New York: HarperCollins Publishers, 1993), p. 3.
58. Hilary F. French, "Assessing Private Capital Flows to Developing Countries," in Lester R. Brown et al., *State of the World 1998* (New York: W.W. Norton & Company, 1998), p. 156.
59. Raveendra N. Batra, *The Myth of Free Trade: A Plan for America's Economic Revival* (New York: Charles Scribner's Sons, 1993), p. 222.
60. Wolfgang Sachs, "Neo-Development: *'Global Ecological Management,'*" in Jerry Mander and Edward Goldsmith, eds., *The Case Against the Global Economy: And for a Turn Toward the Local* (San Francisco: Sierra Club Books, 1996), p. 244.
61. Tom Athanasiou, *Divided Planet: The Ecology of Rich and Poor* (Boston: Little, Brown, 1996), p. 44.
62. *See* Thomas L. Friedman, *The Lexus and the Olive Tree* (New York: Farrar, Straus and Giroux, 1999).
63. John Gray, *False Dawn: The Delusions of Global Capitalism* (New York: The New Press, distributed by W.W. Norton & Company, 1998), p. 3.
64. George Soros, *The Crisis of Global Capitalism: Open Society Endangered* (New York: BBS/PublicAffairs, 1998), p. xi.
65. Ibid., p. 36.
66. Ibid., p. xvi.
67. Daly, *Beyond Growth*, pp. 191–98.
68. Garrett Hardin, *Living Within Limits: Ecology, Economics, and Population Taboos* (New York: Oxford University Press, 1993), pp. 45, 193–94.

Chapter 10. The World Through the Eyes of Thermodynamics

1. Quoted in "Thoughts on the Business of Life," *Forbes*, 12 October 1981, p. 286.
2. Quoted in Molly O'Meara, "One Step Forward, Two Steps Back" (editorial), *World Watch*, May/June 1998, p. 2.
3. R. Buckminster Fuller, *Critical Path* (New York: St. Martin's Press, 1981), p. 133.
4. J. Peter Vajk, *Doomsday Has Been Cancelled* (Culver City, Calif.: Peace Press, 1978), pp. 95–96.
5. Emmett J. Horton and W. Dale Compton, "Technological Trends in Automobiles," *Science*, 10 August 1984, p. 591.
6. O'Meara, "One Step Forward, Two Steps Back," p. 2.
7. Ibid.
8. Mentioned in Gregg Easterbrook, "Greenhouse Common Sense," *U.S. News & World Report*, 1 December 1997, p. 60.
9. "Playboy Interview: George Gilder," *Playboy*, August 1981, p. 86.
10. "A Conversation with Witold Rybczynski: Small May Be Beautiful—But It's Not the Answer," *U.S. News & World Report*, 27 October 1980, p. 53.
11. Aaron Sachs, "Virtual Ecology," *World Watch*, January/February 1999, p. 16.
12. Max Planck, *Treatise on Thermodynamics*, 3rd ed., trans. by Alexander Ogg (New York: Dover Publications, 1945), p. 104.
13. Mathis Wackernagel and William Rees, *Our Ecological Footprint: Reducing Human Impact on the Earth* (Philadelphia, Penn.: New Society Publishers, 1996), p. 129.
14. Gary Gardner and Payal Sampat, "Forging a Sustainable Materials Economy," in Lester R. Brown et al., *State of the World 1999* (New York: W.W. Norton & Company, 1999), p. 51.
15. Edward Tenner, *Why Things Bite Back: Technology and the Revenge of Unintended Consequences* (New York: Alfred A. Knopf, 1996), p. ix.
16. Janet N. Abramovitz and Ashley T. Mattoon, "Reorienting the Forest Products Economy," in Brown et al., *State of the World 1999*, p. 69.
17. Greg Miller, "Law of Unintended Consequences Rules the High-Tech Office," *Los Angeles Times*, 21 November 1999, Business Part II, p. 18.
18. Abramovitz and Mattoon, "Reorienting the Forest Products Economy," pp. 69–70.
19. Population from U.S. Department of Commerce, *Statistical Abstract of the United States, 1998* (Washington, D.C.: U.S. Government Printing Office, 1998), p. 8; materials consumption from Gardner and Sampat, "Forging a Sustainable Materials Economy," p. 46.

20. Lester R. Brown and Christopher Flavin, "A New Economy for a New Century," in Brown et al., *State of the World 1999*, p. 8; Christopher Flavin and Seth Dunn, "Reinventing the Energy System," in ibid., p. 39.
21. Richard P. Turco, *Earth Under Siege: From Air Pollution to Global Change* (New York: Oxford University Press, 1997), p. 401.
22. Joshua Karliner, *The Corporate Planet: Ecology and Politics in the Age of Globalization* (San Francisco: Sierra Club Books, 1997), p. 16.
23. George Gilder, *Wealth and Poverty* (New York: Basic Books, 1981), p. 261.
24. George Gilder, *Microcosm: The Quantum Revolution in Economics and Technology* (New York: Simon and Schuster, 1989), p. 378.
25. Gilder, *Wealth and Poverty*, p. 261.
26. René Dubos, *Reason Awake: Science for Man* (New York: Columbia University Press, 1970), p. 257.
27. Alvin Toffler, *Future Shock* (New York: Random House/Bantam Books, 1970), p. 440.
28. Ibid., p. 486.
29. Ibid., p. 428.
30. Ibid., pp. 486–87.
31. Ibid., p. 428.
32. Alvin Toffler, *The Third Wave* (New York: William Morrow/Bantam Books, 1980), pp. 3–4.
33. Ibid., p. 10.
34. Ibid.
35. Ibid., p. 177.
36. Dubos, *Reason Awake*, p. 257.
37. Toffler, *Future Shock*, p. 429.
38. Ibid.
39. Julius A. Stratton, "The M.I.T. 1964 Commencement Address," in John G. Burke, ed., *The New Technology and Human Values* (Belmont, Calif.: Wadsworth Publishing Company, 1966), p. 94.
40. William M. Bulkeley, "Electronic Gear Aims to Help Folks Cope With Data Deluge," *Wall Street Journal*, 14 August 1981, p. 25.
41. Jacques Ellul, *The Technological Society* (New York: Alfred A. Knopf, 1964; Paris: Librairie Armand Colin, 1954), p. 105.
42. "Editor's Note: If Everyone Has 'Solutions,' Why Are There So Many Problems?" *Informationweek*, 24 February 1986, p. 6.
43. *Business Systems Planning, Information Systems Planning Guide*, IBM Document No. GE20-0527-4, 4th ed., July 1984, p. 69.
44. Craig Brod, *Technostress: The Human Cost of the Computer Revolution* (Reading, Mass.: Addison-Wesley, 1984).

45. Connie Koenenn, "Technostress," *Los Angeles Times*, 11 May 1990, Section E, p. 1.
46. Michael J. Mandel, "You Ain't Seen Nothing Yet," Special Double Issue; The 21st Century Economy, *Business Week*, 24–31 August 1998, pp. 60–63.
47. "A Conversation with Paul B. MacCready, Jr., 'Unorthodox Approach Needed' to Solve Today's Problems," *U.S. News & World Report*, 2 August 1982, p. 69.
48. Ibid.
49. Lindsey Grant, *Juggernaut: Growth on a Finite Planet* (Santa Ana, Calif.: Seven Locks Press, 1996), p. 92.
50. Quoted in "Thoughts on the Business of Life," *Forbes*, 12 October 1981, p. 286.
51. "Calvin Coolidge's Bum Rap," *Newsweek*, 15 August 1983, p. 22.
52. Marvin Stone, "The Good Sense of 'Silent Cal,'" *U.S. News & World Report*, 6 July 1981, p. 72.
53. "Calvin Coolidge's Bum Rap," p. 22.
54. Stone, "The Good Sense of 'Silent Cal,'" p. 72.
55. "Yes, Clean Living Does Pay Off," *U.S. News & World Report*, 10 November 1975, p. 58.
56. Nancy Gibbs, "Shameful Bequests to the Next Generation," *Time*, 8 October 1990, p. 45.
57. Earl Cook, *Man, Energy, Society* (San Francisco: W. H. Freeman, 1976), p. 157.
58. G. Tyler Miller, Jr., *Energetics, Kinetics, and Life: An Ecological Approach* (Belmont, Calif.: Wadsworth Publishing Company, 1971), p. 293.
59. "Matters of Scale: Visible vs. Invisible Waste," *World Watch*, November/December 1999, p. 37.
60. Janet Luhrs, *The Simple Living Guide: A Sourcebook for Less Stressful, More Joyful Living* (New York: Broadway Books, 1997), p. 248.
61. U.S. Department of Agriculture, Study Team on Organic Farming, *Report and Recommendations on Organic Farming* (Washington, D.C.: July 1980), p. xiii; "Nature vs. Nurture on the Farm," *U.S. News & World Report*, 18 September 1989, p. 53.
62. Mark L. Winston, *Nature Wars: People vs. Pests* (Cambridge, Mass.: Harvard University Press, 1997), p. 186.
63. Ibid., p. 44.
64. Ibid., pp. 44–45.
65. Ibid., p. 186.
66. Edward Teller, *Energy from Heaven and Earth* (San Francisco: W. H. Freeman, 1979), p. 85.
67. Vance Packard, *The Waste Makers* (New York: David McKay Company, 1960), pp. 53–67.

68. "A Conversation with Marcus Cunliffe: Behind America's 'Excessive Fascination with Novelty,'" *U.S. News & World Report*, 7 September 1981, p. 66.

69. Donald B. Kraybill, *The Riddle of Amish Culture* (Baltimore: The Johns Hopkins University Press, 1989), p. 43.

70. Ibid., p. 44.

71. Ibid., p. 45.

72. Marshall Sahlins, *Stone Age Economics* (Chicago: Aldine-Atherton, 1972), pp. 34–35, 37. *See also* Bruce Rich, *Mortgaging the Earth: The World Bank, Environmental Impoverishment, and the Crisis of Development* (Boston: Beacon Press, 1994), pp. 202–3.

73. Richard Folkers, "Returning to Retro-tech," *U.S. News & World Report*, 1 December 1997, pp. 93–97. *See also* Merrill Markoe, "I Am Not a Technochicken," in ibid., p. 92.

74. Christopher Evans, *The Micro Millennium* (New York: Viking Press/Washington Square Press, 1980, copyright 1979), p. 68.

75. Edward Teller, *The Pursuit of Simplicity* (Malibu, Calif.: Pepperdine University Press, 1980), p. 113; Robert Jastrow, *The Enchanted Loom: The Mind in the Universe* (New York: Simon and Schuster, 1981), p. 162.

76. Duane Elgin, *Voluntary Simplicity: Toward a Way of Life That Is Outwardly Simple, Inwardly Rich*, rev. ed. (New York: William Morrow/Quill, 1993), p. 187.

77. Ibid.

78. Sachs, "Virtual Ecology," p. 13.

79. Ibid., pp. 14, 18.

Chapter 11. The Thermodynamic Imperative

1. H. S. Seifert, "Can We Decrease Our Entropy?" *American Scientist*, Summer-June 1961, p. 124A.

2. Leo Tolstoy, "The Superstitions of Science," *The Arena*, 20, 1898, pp. 52–60, reprinted in John G. Burke, ed., *The New Technology and Human Values* (Belmont, Calif.: Wadsworth Publishing Company, 1966), pp. 27–28.

3. Ibid., p. 26.

4. Ibid., p. 30.

5. R. B. Lindsay, "Entropy Consumption and Values in Physical Science," *American Scientist*, Autumn-September 1959, p. 384.

6. Seifert, "Can We Decrease Our Entropy?" p. 128A.

7. G. Tyler Miller, Jr., *Energetics, Kinetics, and Life: An Ecological Approach* (Belmont, Calif.: Wadsworth Publishing Company, 1971), p. 330.

8. Ibid.

9. Pierre Teilhard de Chardin, *The Phenomenon of Man* (New York: Harper

     & Brothers, 1959), p. 249.
10. C. P. Snow, *The Physicists* (Boston: Little, Brown, 1981), p. 16.
11. Steven Weinberg, "Five and a Half Utopias," *Atlantic Monthly,* January
     2000, p. 113.
12. René Dubos, *Reason Awake: Science for Man* (New York: Columbia Uni-
     versity Press, 1970), p. 216.
13. Ibid., pp. 216–17.
14. Paul A. Samuelson, *Economics*, 11th ed. (New York: McGraw-Hill, 1980),
     p. 747.
15. Dubos, *Reason Awake*, pp. 178–79.
16. Ibid., p. 198.
17. René Dubos, *Celebrations of Life* (New York: McGraw-Hill, 1981), p.
     236.
18. Sharon Begley, "On the Wings of Icarus," *Newsweek*, 20 May 1991,
     p. 64.
19. Ibid., pp. 64–65.
20. Ibid., p. 65.
21. Ibid.
22. David W. Ehrenfeld, *The Arrogance of Humanism* (New York: Oxford
     University Press, 1978), p. 119.
23. Matthew L. Wald, "E.P.A. Says Catalytic Converter Is Growing Cause
     of Global Warming," *New York Times*, 29 May 1998, pp. A1, A16.
24. Quoted in Barry Commoner, *The Closing Circle: Nature, Man, and Tech-
     nology* (New York: Alfred A. Knopf/Bantam Books, 1971), p. 179.
25. Max Planck, *Treatise on Thermodynamics*, 3rd ed., trans. by Alexander
     Ogg (New York: Dover Publications, 1945), p. 84.
26. Dan Krause, "Letters: Muddy Mississippi Waters," *Newsweek*, 7 May
     1990, p. 12.
27. "Not Just for Nerds," *Newsweek*, 9 April 1990, pp. 55, 62.
28. Ibid., pp. 62–63.
29. Stephan Wilkinson, "The Automobile and the Environment: Our Next
     Car?" *Audubon*, May-June 1993, p. 58.
30. Betsy Carpenter, "Living with Our Legacy," *U.S. News & World Report*,
     23 April 1990, p. 65.
31. Dennis Overbye, "Caution: We Brake for Newton," *Time*, 7 October
     1991, p. 74.
32. Ibid.
33. Allan Bloom, *The Closing of the American Mind* (New York: Simon and
     Schuster, 1987), p. 380.
34. Ibid., p. 339.
35. Ibid., pp. 344–47.
36. Ibid., p. 344.

37. Julian Huxley, "Science and Synthesis," in *Science and Synthesis: An International Colloquium Organized by Unesco on the Tenth Anniversary of the Death of Albert Einstein and Teilhard de Chardin* (New York: Springer-Verlag, 1971, copyright 1967), p. 29.
38. Julius A. Stratton, "The M.I.T. 1964 Commencement Address," in Burke, *New Technology and Human Values*, p. 95.
39. Ibid.
40. Ibid.
41. Huxley, "Science and Synthesis," p. 29.
42. Mentioned in Ferdinand Gonseth, "Einstein's Knowledge of Nature and Philosophy," in *Science and Synthesis*, p. 7.
43. Quoted in José M. R. Delgado, M.D., *Physical Control of the Mind: Toward a Psychocivilized Society.* Vol. 41, *World Perspectives* (New York: Harper & Row, 1969), p. 13.
44. William Thomson (Lord Kelvin), "On an Absolute Thermometric Scale Founded on Carnot's Theory of the Motive Power of Heat, and Calculated from Regnault's Observations," in Joseph Kestin, ed., *The Second Law of Thermodynamics* (Stroudsburg, Penn.: Dowden, Hutchinson & Ross, 1976), p. 54.
45. Huxley, "Science and Synthesis," p. 30.
46. Edward O. Wilson, *Consilience: The Unity of Knowledge* (New York: Alfred A. Knopf, 1998), pp. 12–13.
47. Ibid., pp. 11–12.
48. Roger Penrose, *The Emperor's New Mind: Concerning Computers, Minds, and the Laws of Physics* (New York: Penguin Books, 1991), p. 304.
49. Arnold Sommerfeld, *Thermodynamics and Statistical Mechanics, Lectures on Theoretical Physics*, Vol. V, trans. by J. Kestin (New York: Academic Press, 1956), p. v.
50. David Bohm, *Quantum Theory* (Englewood Cliffs, N.J.: Prentice-Hall, 1951), pp. 608–9.
51. Harold J. Morowitz, *Entropy for Biologists: An Introduction to Thermodynamics* (New York: Academic Press, 1970); Raymond Kern and Alain Weishrod, *Thermodynamics for Geologists*, trans. by Duncan McKie (San Francisco: Freeman, Cooper, 1967); Nicholas Georgescu-Roegen, *The Entropy Law and the Economic Process* (Cambridge, Mass.: Harvard University Press, 1971); P. T. Landsberg, *Entropy and the Unity of Knowledge* (Cardiff: University of Wales Press, 1964).
52. *See* C. E. Shannon and W. Weaver, *The Mathematical Theory of Communication* (Urbana, Ill.: University of Illinois Press, 1964, copyright 1949), and Norbert Wiener, *Cybernetics, or Control and Communication in the Animal and the Machine*, 2nd ed. (New York: The MIT Press and John Wiley, 1961).

53. Wilson, *Consilience*, p. 285. *See also* Traci Watson, "Global Warming Is Accelerating, Scientists Report," *USA Today*, 26 October 2000, p. 1A.

54. Max Planck, *Scientific Autobiography, and Other Papers*, trans. by Frank Gaynor (New York: Greenwood Press, 1968, copyright 1949 by Philosophical Library), pp. 19–20.

55. Ibid., pp. 37–38.

56. Abraham Pais, *Niels Bohr's Times: In Physics, Philosophy, and Polity* (New York: Oxford University Press, 1991), pp. 86–87.

57. Quoted in Walter Kaufmann, *The Future of the Humanities* (New York: Reader's Digest Press, distributed by Thomas Y. Crowell Company, 1977), p. 2.

58. Ibid., p. 3.

59. Brian Greene, *The Elegant Universe: Superstrings, Hidden Dimensions, and the Quest for the Ultimate Theory* (New York: W.W. Norton & Company, 1999), p. 334.

60. Ibid., p. 337.

61. Frederick Turner, "Design for a New Academy: An End to Division by Department," *Harper's Magazine*, September 1986, p. 47.

62. Quoted in Richard Feynman, *The Character of Physical Law* (Cambridge, Mass.: The MIT Press, 1967), p. 58.

63. Turner, "Design for a New Academy," p. 50.

Index